AS/A-LEVEL YEAR 1

STUDENT GUIDE

EDEXCEL

Chemistry

Topics 1–5

Atomic structure and the periodic table
Bonding and structure
Redox I
Inorganic chemistry and the periodic table
Formulae, equations and amounts of substance

George Facer

Rod Beavon

D1493084

PHILIP ALLAN FOR
HODDER
EDUCATION
AN HACHETTE UK COMPANY

Philip Allan, an imprint of Hodder Education, an Hachette UK company, Blenheim Court, George Street, Banbury, Oxfordshire OX16 5BH

Orders

Bookpoint Ltd, 130 Milton Park, Abingdon, Oxfordshire OX14 4SB

tel: 01235 827827

fax: 01235 400401

e-mail: education@bookpoint.co.uk

Lines are open 9.00 a.m.–5.00 p.m., Monday to Saturday, with a 24-hour message answering service. You can also order through the Hodder Education website: www.hoddereducation.co.uk

© George Facer and Rod Beavon 2015

ISBN 978-1-4718-4333-4

First printed 2015

Impression number 5 4 3 2 1

Year 2019 2018 2017 2016 2015

This guide has been written specifically to support students preparing for the Edexcel AS and A Level Chemistry examinations. The content has been neither approved nor endorsed by Edexcel and remains the sole responsibility of the authors.

Cover photo: TTstudio/Fotolia

Typeset by Integra Software Services Pvt. Ltd, Pondicherry, India

Printed in Italy

Hachette UK's policy is to use papers that are natural, renewable and recyclable products and made from wood grown in sustainable forests. The logging and manufacturing processes are expected to conform to the environmental regulations of the country of origin.

Contents

Getting the most from this book 4

About this book 5

Content Guidance

Topic 1 Atomic structure and the periodic table 8

Topic 2 Bonding and structure 14

    Topic 2A Bonding. 14

    Topic 2B Structure 26

Topic 3 Redox I 29

Topic 4 Inorganic chemistry and the periodic table 32

    Topic 4A Group 2 (alkaline earth metals). 32

    Topic 4B Group 7 (halogens) 36

    Topic 4C Analysis of inorganic compounds 41

Topic 5 Formulae, equations and amounts of substance 42

    Core practical 1 Measure the molar volume of a gas

Practical aspects 56

    Core practical 2 Prepare a standard solution from a solid acid and
    use it to find the concentration of a solution of sodium hydroxide

    Core practical 3 Find the concentration of a solution of
    hydrochloric acid

Questions & Answers

Structured and multiple-choice questions 63

Knowledge check answers 80

Index 82

Periodic table 85

■ Getting the most from this book

Exam tips

Advice on key points in the text to help you learn and recall content, avoid pitfalls, and polish your exam technique in order to boost your grade.

Knowledge check

Rapid-fire questions throughout the Content Guidance section to check your understanding.

Knowledge check answers

1 Turn to the back of the book for the Knowledge check answers.

Summaries

■ Each core topic is rounded off by a bullet-list summary for quick-check reference of what you need to know.

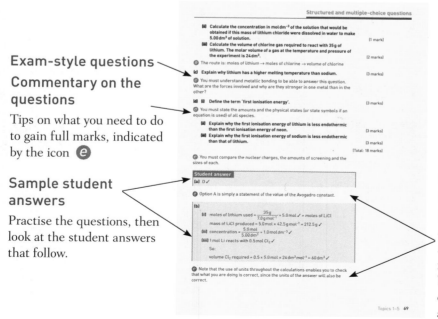

Exam-style questions

Commentary on the questions

Tips on what you need to do to gain full marks, indicated by the icon ⓔ

Sample student answers

Practise the questions, then look at the student answers that follow.

Commentary on sample student answers

Find out how many marks each answer would be awarded in the exam and then read the comments (preceded by the icon ⓔ) showing exactly how and where marks are gained or lost.

■ About this book

This guide is the first of a series covering the Edexcel specification for AS and A Level Chemistry. It offers advice for the effective study of Topics 1 to 5, which are examined on AS paper 1 and on A Level papers 1, 2 and 3. The aim of this guide is to help you understand the chemistry — it is not intended as a shopping list, enabling you to cram for an examination. The guide has two sections:

■ The **Content Guidance** is not intended to be a textbook. It offers guidelines on the main features of the content of Topics 1 to 5, together with advice on making study more productive.

■ The **Questions & Answers** section gives advice on approaches and techniques to ensure you answer the examination questions in the best way you can. It also provides exam-style questions, with student answers and comments on how the answers would be marked.

The effective understanding of chemistry requires time. No one suggests that chemistry is an easy subject, but if you find it difficult you can overcome your problems by the proper investment of time.

To understand the chemistry, you have to make links between the various topics. The subject is coherent and is not a collection of discrete modules. Once you have spent time thinking about chemistry, working with it and solving chemical problems, you will become aware of these links. Spending time this way will make you fluent with the ideas. Once you have that fluency, and practise the good techniques described in this book, the examination will look after itself. Don't be an examination automaton — be a chemist.

The specification

The specification states the chemistry that can be tested in the examinations and the format of those exams. It can be obtained from Edexcel, either as a printed document or from the web at www.edexcel.com.

Learning to learn

Learning is not instinctive — you have to develop suitable techniques to make effective use of your time. In particular, chemistry has peculiar difficulties that need to be understood if your studies are to be effective from the start.

Planning

Efficient people do not achieve what they do by approaching life haphazardly. They plan — so that if they are working, they mean to be working, and if they are watching television, they have planned to do so. Planning is essential. You must know what you have to do each day and set aside time to do it. Furthermore, to devote time to study means you may have to give something up that you are already doing. There is no way that you can generate extra hours in the day.

Be realistic in your planning. You cannot work all the time, and you must build in time for recreation and family responsibilities.

Targets

When devising your plan, have a target for each study period. This might be a particular section of the specification, or it might be rearranging information from text into pictures, or drawing a flowchart relating all the reactions of groups 1 and 2. Whatever it is, be determined to master your target material before you leave it.

Reading chemistry textbooks

A page of chemistry may contain a range of material that differs widely in difficulty. Therefore, the speed at which the various parts of a page can be read may have to vary, if it is to be understood. You should read with pencil and paper to hand and jot things down as you go — for example, equations, diagrams and questions to be followed up. If you do not write down the questions, you will forget them; if you do not master detail, you will never become fluent in chemistry.

Practising skills

Chemical equations

Equations are used because they are quantitative, concise and internationally understood. Take time over them, copy them and check that they balance. Most of all, try to visualise what is happening as the reaction proceeds. If you cannot, make a note to ask someone who can or — even better — ask your teacher to *show* you the reaction if at all possible. Chemical equations describe real processes; they are not abstract algebraic constructs.

Graphs

Graphs give a lot of information, and they must be understood in detail rather than as a general impression. Take time over them. Note what the axes are, what the units are, the shape of the graph and what the shape means in chemical terms.

Tables

Tables are a means of displaying a lot of information. You need to be aware of the table headings and the units of numerical entries. Take time over them. What trends can be seen? How do these relate to chemical properties? Sometimes it can be useful to convert tables of data into graphs.

Diagrams

Diagrams of apparatus should be drawn in section. When you see them, copy them and ask yourself why the apparatus has the features that it has. What is the difference between distillation and reflux apparatus, for example? When you do practical work, examine each piece of the apparatus closely so that you know both its form and function.

Calculations

Do not take calculations on trust — work through them. First, make certain that you understand the problem, and then that you understand each step in the solution.

Make clear the units of the physical quantities used and make sure you understand the underlying chemistry. If you have problems, ask.

Always make a note of problems and questions that you need to ask your teacher. Learning is not a contest or a trial. Nobody has ever learnt anything without effort or without running into difficulties from time to time — not even your teachers.

Notes

Most people have notes of some sort. Notes can take many forms: they might be permanent or temporary; they might be lists, diagrams or flowcharts. You have to develop your own styles. For example, notes that are largely words can often be recast into charts or pictures; this is useful for imprinting the material. The more you rework the material, the clearer it will become.

Whatever form your notes take, they must be organised. Notes that are not indexed or filed properly are useless, as are notes written at enormous length and those written so cryptically that they are unintelligible a month later.

Writing

In chemistry, extended writing is often not required. However, you need to be able to write concisely and accurately. This requires you to marshal your thoughts properly and needs to be practised during your day-to-day learning.

Have your ideas assembled in your head before you start to write. You might imagine them as a list of bullet points. Before you write, have an idea of how you are going to link these points together and also how your answer will end. The space available for an answer is a poor guide to the amount that you have to write — handwriting sizes differ hugely, as does the ability to write succinctly. Filling the space does not necessarily mean you have answered the question. The mark allocation suggests the number of points to be made, not the amount of writing needed.

Content Guidance

■ Topic 1 Atomic structure and the periodic table

Atoms have a central nucleus containing protons and neutrons. The diameter of the nucleus is about 10^{-5} of the atomic diameter. The electrons are located in shells of different energy levels surrounding the nucleus. The shells are made up of orbitals. The important properties of these three particles are given in Table 1.

	Mass relative to proton	Charge relative to proton
Proton	1	+1
Neutron	1.001	0
Electron	$\dfrac{1}{1836}$	−1

Table 1

Definitions

- **Atom** — the smallest *uncharged* part of an element.
- **Atomic number** — the number of protons in an atom. This is written as a subscript number to the left of the element's symbol, for example $_{11}Na$.
- **Mass number** — the number of nucleons (protons plus neutrons) in an atom. This is written as a superscript number to the left of the element's symbol, for example ^{23}Na.
- **Isotopes** — atoms having the same proton number but different mass numbers. They have the same electronic structure and therefore the same chemistry. They differ only in mass.
- **Relative isotopic mass** — the mass of a particular isotope of an element, relative to $\frac{1}{12}$ the mass of a carbon-12 atom.
- **Relative atomic mass, A_r** — the weighted mean of the masses of the atoms of the element relative to $\frac{1}{12}$ the mass of a carbon-12 atom. For chlorine, which is 75% ^{35}Cl and 25% ^{37}Cl, the relative atomic mass is $(0.75 \times 35) + (0.25 \times 37) = 35.5$. This number is given in the periodic table at the back of the book.
- **Relative molecular mass, M_r** — the mass of a molecule relative to $\frac{1}{12}$ the mass of a carbon-12 atom.
- **Relative formula mass** — the sum of all the relative atomic masses of all the atoms in a given formula. This term is used for ionic substances and those with giant structures, such as silicon dioxide.

Knowledge check 1

Two isotopes of bromine exist: $^{79}_{35}Br$ and $^{81}_{35}Br$. Deduce the number of protons and neutrons in each isotope.

Knowledge check 2

Calculate the relative formula mass of calcium hydroxide, $Ca(OH)_2$.

The mass spectrometer

The mass spectrometer can be used to determine relative atomic mass. It works as follows:

- Gaseous samples of elements or compounds are bombarded with high-energy electrons.
- This causes ionisation, producing positive ions.
- These ions are accelerated by an electric field.
- They are then bent by a magnetic field into a circular path whose radius depends on their mass.
- A detector enables the number of ions at each particular mass to be determined.

The resulting mass spectrum gives information about the isotopic composition of elements or structural information about compounds.

The mass spectrum of ethanol, CH_3CH_2OH, is given in Figure 1.

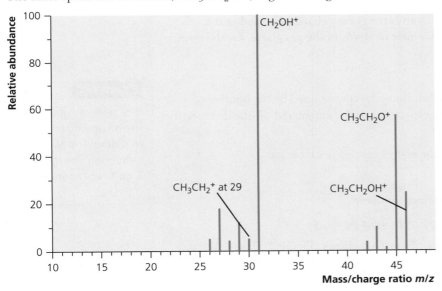

Figure 1 The mass spectrum of ethanol

The largest value of m/z equals the value of the relative molecular mass of ethanol, which is 46. The other lines represent how ethanol molecules fragment.

Uses of mass spectral data include:

- determination of the isotopic composition of an element. This was first achieved for neon, when F. W. Aston showed the presence of ^{10}Ne and ^{11}Ne, thus discovering that isotopes exist.

> **Exam tip**
>
> The relative peak heights depend on the isotopic abundances. The two isotopes of bromine are in equal proportions, so the chances of ^{79}Br–^{79}Br and ^{81}Br–^{81}Br are ½ × ½ = ¼ for each. The chances for ^{79}Br–^{81}Br are 2 × ½ × ½ = ½. Thus the heights of the peaks will be in the ratio of 1:2:1.

> **Exam tip**
>
> Do not forget the *positive* charge when giving the formula of a species causing a peak in a mass spectrum.

> **Exam tip**
>
> Not all compounds give a molecular ion peak, as the M^+ ion may be too unstable.

> **Knowledge check 3**
>
> Bromine has two stable isotopes, ^{79}Br and ^{81}Br. Explain how many peaks due to the Br_2^+ ion will be seen in its mass spectrum.

■ deduction of the relative atomic mass of an element. This involves finding the masses and abundances of the isotopes.

■ deduction of the structure of an organic compound from the *m/z* values of the molecular ion and the fragments that arise from bombardment of the molecules.

■ measurement of the $^{14}C:^{12}C$ ratio in the radioactive dating of organic material. The older the material, the smaller this ratio is. This application (known as carbon dating) showed that the flax used to make the Turin Shroud, supposedly the burial shroud of Jesus, was grown within a few years of 1305.

■ identification of molecules in blood and urine. This application is used on samples taken from athletes for drug testing.

Ionisation energy and electron configuration

Ionisation energy

The **first ionisation energy** for an atom is the energy change involved in the removal of one electron from each of a mole of atoms in the gas phase, i.e. the energy change per mole for:

$$E(g) \rightarrow E^+(g) + e^-$$

The value of the ionisation energy of an element is determined by the number of protons, the shielding effect of inner electrons and the atomic radius (the distance the outer electron is from the nucleus).

The **second ionisation energy** is the energy change per mole for:

$$E^+(g) \rightarrow E^{2+}(g) + e^-$$

Successive ionisation energies are defined similarly.

Trend in first ionisation energy across a period

The general trend is an increase in first ionisation energy (Figure 2). This is because there is a greater attraction between the outer electrons and the nucleus, which is caused by:

■ an increase in the number of protons

■ the outer electrons being shielded to approximately the same extent by the inner electrons

■ the atomic radius being smaller

There are two discontinuities. One is between groups 2 and 3, where the outer electron is now in the higher energy *p*-orbital. Another discontinuity is between groups 5 and 6, where the electron removed is from the *p*-orbital and is repelled by the other electron in that orbital.

Trend in first ionisation energy down a group

The first ionisation energies decrease down a group (Figure 2). The number of protons increases, but so does the number of shielding electrons. However, the outer electrons are in a shell further from the nucleus and so are attracted less strongly.

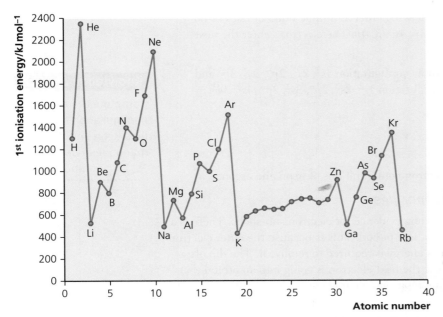

Figure 2 First ionisation energies of the first 37 elements of the periodic table

Electron configuration

Spectral evidence

When hydrogen is energised in a discharge tube, spectral lines of different frequencies are emitted. One series, the Lyman series, is found in the ultraviolet region and another, the Balmer series, in the visible part of the electromagnetic spectrum. Each series has lines with frequencies that converge on a single value.

When hydrogen atoms are heated, electrons are promoted from the **ground state** (their most stable or lowest energy state) to a higher, excited state. In terms of electron orbits, this means that the electron is promoted from the first orbit to an outer orbit. This is not a stable state and the electron drops back, giving out energy in the form of light. The energy given out is the difference between the energy of the electron in its outer orbit and the energy of the orbit into which it drops.

- The Lyman series is obtained when an electron in a hydrogen atom drops from an excited state back to the ground state, which is when it is in the first orbit.
- The Balmer series is caused by the electron dropping back to the second orbit.

Successive ionisation energies

Successive ionisation energies show the existence of quantum shells; there are significant jumps in ionisation energy between each shell. Thus, for aluminium there is a big jump between the third and fourth ionisation energies, as the fourth electron has to be removed from the inner second shell.

Elements of groups 1 and 2 have *s*-electrons as their outer electrons and constitute the *s*-block; those of groups 3 to 0 have outer *p*-electrons and constitute the *p*-block. In the *d*-block, the *d*-sub-shell is being filled — for example, $_{21}$Sc to $_{30}$Zn are *d*-block elements. However, the outer electrons for these elements are *s*-electrons.

The electron energy levels fill in the following order as atomic number rises as far as krypton (element 36): 1s, 2s, 2p, 3s, 3p, 4s, 3d, 4p. The electrons enter the next sub-shell, once the previous one is full.

Thus, phosphorus, $_{15}$P, has the electronic configuration $1s^2, 2s^2, 2p^6, 3s^2, 3p^3$ and vanadium, $_{23}$V, has the electronic configuration $1s^2, 2s^2, 2p^6, 3s^2, 3p^6, 4s^2, 3d^3$. These are often written as:

for phosphorus   [Ne] $3s^2 3p^3$

for vanadium   [Ar] $4s^2 3d^3$

where [Ne] and [Ar] represent the electron configurations of neon and argon.

Evidence from first ionisation energies

The repeating pattern in Figure 2 is strong evidence for electron sub-shells. There is a significant fall after the second element in a period. This is because the outer electron is in a higher energy orbital and so less energy is required to remove it. The dip after the fifth element in a period is because the next electron is going into an orbital that already has one electron. The resulting repulsion means that less energy is required to remove it.

Orbitals

Chemistry is concerned with elements bonding together, which involves electrons. The electronic structure therefore determines the chemistry of an element.

The electrons in atoms do not orbit the nucleus like planets around the Sun. They exist as volumes of electron density — the **orbitals** — centred on the nucleus (Figure 3).

Section through s-orbital   One p-orbital

Figure 3 (a) Section through an s-orbital; (b) one p-orbital

The p sub-shells are divided into three different orbitals, the p_x, the p_y and the p_z orbitals. These point along the x, y and z axes. The d sub-shells are divided into five orbitals.

In both the p and the d sub-shells, the electrons are unpaired as far as possible. This is shown by the 'electrons-in-boxes' notation, for example Figure 4.

Phosphorus [Ne], 3s $\boxed{\uparrow\downarrow}$ $3p_x$ $\boxed{\uparrow}$ $3p_y$ $\boxed{\uparrow}$ $3p_z$ $\boxed{\uparrow}$

Figure 4 Electrons-in-boxes diagram for phosphorus

Covalent bonds are formed by the overlap of orbitals on different atoms, usually one electron each, to give a molecular orbital that can be represented by an electron-density map. This maps the density from the pair of electrons that form the bond. The s-orbital is a sphere centred on the nucleus; each of the three p-orbitals (one along each axis) is rather like two balloons at 180°.

Knowledge check 5

Complete the electron configuration of (a) silicon and (b) manganese using the 1s, 2s... notation.

Exam tip

An orbital can hold a maximum of two electrons.

Detailed electron configuration

- The electrons fill up in the order:
  $1s, 2s\ 2p, 3s\ 3p, 4s, 3d\ldots$
- An orbital can hold up to two electrons.
- When there are two electrons in an orbital, they must have different spins.
- The one s-orbital in a shell can hold up to two electrons.
- The three p-orbitals in a shell can hold up to six electrons in total.
- The five d-orbitals in a shell can hold up to ten electrons in total.
- In a given sub-shell (e.g. the $2p$) the electrons remain as unpaired as possible. Thus carbon is $1s^2\ 2s^2\ 2p_x^1\ 2p_y^1$ not $1s^2\ 2s^2\ 2p_x^2$.

There are two discrepancies among the first 36 elements:

- Chromium is $[Ar]\ 4s^1\ 3d^5$ not $[Ar]\ 4s^2\ 3d^4$.
- Copper is $[Ar]\ 4s^1\ 3d^{10}$ not $[Ar]\ 4s^2\ 3d^9$.

This is due to the extra stability of a half-filled or totally filled set of d-orbitals.

The chemical properties of an element are determined by its electronic configuration. Thus carbon, $[Ne]\ 2s^2\ 2p^2$, has four electrons in its outer orbit and so can form four covalent bonds (see page 16).

Elements that have their highest energy electron in an s-orbital are described as being in the s block of the periodic table. These are the groups 1 and 2 elements.

Those with their highest energy electron in a p-orbital are described as being in the p block. These are elements in groups 3 to 7 plus the noble gases (group 0).

Periodic properties

Periodicity is the repeating pattern of physical or chemical properties across different periods.

The reason that elements in the same group have similar properties is that some properties recur periodically. Ionisation energy is one example, as shown in the graph for period 2 (elements 3–10) and period 3 (elements 11–18) (Figure 2 — see page 11). Thus, all the alkali metals have low ionisation energies; the noble (inert) gases have high ionisation energies.

The melting and boiling temperatures of the elements of periods 2 and 3 also show periodic trends. The values for the metals in groups 1–3, with strong metallic bonding, are much higher than those for the non-metals in groups 5–0, but are comparable to those of the giant covalent structures of the group 4 elements carbon and silicon, which have many covalent bonds between the atoms and so a very high melting temperature.

The melting and boiling temperatures of an element reflect the binding energies in the crystal or liquid states. In metals, this depends on the size of the ions, their charge and the number of delocalised electrons. The metallic radii in period 3 are larger than those in period 2, so the metals have lower melting and boiling temperatures. With the elements in groups 5–0, the more electrons in the molecule, the greater are the London forces (see page 23) and the higher the melting and boiling temperatures.

Exam tip

The $4s$ electrons can be written either before those of the $3d$ (in order of energy) or after (in order of distance from the nucleus).

Knowledge check 6

Complete the electrons-in-boxes diagram for vanadium (atomic number 23):

[Ar]...

Exam tip

Remember that atoms get *smaller* across a period and *bigger* down a group.

Periodic trends are shown in Table 2.

Element	Li	Be	B	C (graphite)	N	O	F	Ne
Melting temp./°C	181	1278	2300	3697	−210	−218	−220	−248
Boiling temp./°C	1342	2970	2550	4827	−196	−183	−188	−246
Structure	Body-centred cubic	Hexagonal close packing	Tetrahedral	Giant molecule	N_2	O_2	F_2	Atoms

Element	Na	Mg	Al	Si	P	S	Cl	Ar
Melting temp./°C	98	649	660	1410	44	119	−101	−189
Boiling temp./°C	883	1107	2467	2355	280	445	−35	−186
Structure	Body-centred cubic	Hexagonal close packing	Face-centred cubic	Giant molecule	P_4 (white phosphorus)	S_8	Cl_2	Atoms

Table 2 Periodic trends

Summary

After studying this topic, you should be able to:
- define atomic and mass numbers
- define isotopes
- define relative atomic and isotopic masses
- define ionisation energy
- define electron affinity
- explain how a mass spectrometer works
- deduce the relative atomic mass from mass spectra data
- state the electron configuration of elements up to $_{38}$Sr
- explain the trends in first ionisation energies across a period and in a group
- explain the trends of melting temperatures of elements in a period

■ Topic 2 Bonding and structure

Topic 2A Bonding

Ionic bonding

An **ionic bond** is the strong electrostatic force of attraction between a positive and a negative ion.

The formation of ions

- Positive ions or **cations** are formed by loss of electron(s) from atoms or groups of atoms. The ions are usually metal ions, but can be non-metallic — for example, the ammonium ion, NH_4^+.
- Negative ions or **anions** are formed when atoms or groups of atoms gain electrons.
- The ions formed may have octets of electrons, but many d-block metal ions do not.
- The driving force for ion formation lies in the lower energy of the compound formed, compared with the constituent elements.

Dot-and-cross diagrams of ions

The electrons in an ionic compound can be shown by having electrons theoretically from one atom shown as dots and those from the other atom as crosses. It is usual to show only the outer electrons in each ion.

For sodium chloride each sodium atom, 2,8,1 gives one electron to each chlorine atom, 2,8,7 (Figure 5). Both reach a noble gas electronic configuration.

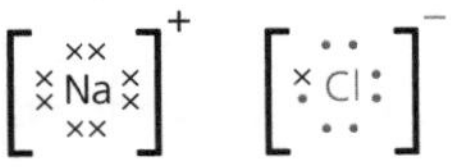

Figure 5 Dot-and-cross diagram for sodium and chloride ions

Ionic radii

Metal ions (cations), which are formed by electron loss, are smaller than the metal atoms from which they arose. This is because, in forming an ion, the metal atom loses its outer electron shell.

A non-metal ion formed from a single atom is slightly larger than the atom, as the addition of extra electron(s) is to the outer shell. However, the extra repulsion between the outer electrons causes the anion to be bigger than the atom. Polyatomic anions, such as sulfate $SO_4{}^{2-}$, are much larger.

The size of ions increases going down a group in the periodic table. Successive members of the group have one more shell of electrons. They also have more protons in the nucleus. The effect of more shells outweighs the greater attraction from the nucleus, so size does increase, but not as dramatically as might be expected (see Table 3).

Group 1	Li	Na	K	Rb	Cs
Atomic radius/nm	0.157	0.191	0.235	0.250	0.272
Ionic radius (+1 ion)/nm	0.074	0.102	0.138	0.149	0.150

Group 7	F	Cl	Br	I
Atomic radius/nm	0.155	0.180	0.190	0.195
Ionic radius (–1 ion)/nm	0.133	0.180	0.195	0.215

Table 3

Isoelectronic ions are ions that have the same electron configuration.

Because their nuclei are different, they are different ions. There are six ions that have the electron configuration $1s^2\,2s^2\,2p^6$. Their radii (in nm) are shown in Table 4. The decrease in ionic radius from nitrogen to aluminium is due to the increasing positive charge on the nucleus, which pulls the electron shells closer in.

Ion	N^{3-}	O^{2-}	F^-	Na^+	Mg^{2+}	Al^{3+}
Nuclear charge	+7	+8	+9	+11	+12	+13
Radius/nm	0.171	0.140	0.133	0.102	0.072	0.053

Table 4

Evidence for the existence of ions

The evidence for the existence of ions in compounds comes from two sources.

Electrolysis

The passage of a current through a molten salt or aqueous salt solution relies on the movement of the ions, which then lose electrons at the anode or gain them at the cathode to form the constituent elements.

Knowledge check 7

Draw the electronic configuration, showing all electrons, and the charge on (a) a magnesium ion and (b) a chloride ion.

Knowledge check 8

Explain why an O^{2-} ion is bigger than an F^- ion.

This movement can be seen if the ions are coloured. Copper(II) chromate, $CuCrO_4$, is green as it contains blue Cu^{2+} and yellow CrO_4^{2-} ions. If a small crystal is placed at the middle of a strip of wetted filter paper attached at each end to a power supply, a yellow colour spreads to the positive electrode and a blue colour to the negative electrode.

Melting temperature

The strong electrostatic force of attraction between ions means that a large amount of energy is required to separate the ions. The result is a high melting temperature. The larger the charge and the smaller the ions, the higher is the melting temperature. Thus aluminium oxide melts at 2072°C, whereas sodium chloride melts at 801°C.

Covalent bonding

A **covalent bond** is the strong electrostatic attraction between two nuclei and a shared pair of electrons. Covalent bonds are strong and have a number of features:

- Electrons are shared in pairs.
- A bond is formed by the overlap of two orbitals.
- Two or three electron pairs can be shared between a pair of atoms to give a double or triple bond.
- The bonded atoms often contain octets but some have only six electrons in the valence (outer) shell — for example, in BCl_3 there are only six electrons round the boron atom. Atoms in period 3 and beyond can have more than eight electrons since *d*-orbitals are available — for example, S in SF_6 has 12 electrons in the valence shell.
- Dative covalent bonds are formed when both bonding electrons come from the same atom.
- Dative bonds are no different from any other covalent bond.
- Dative bonds are found:
  - in NH_4^+
  - between chlorine and electron-deficient aluminium in Al_2Cl_6
  - as the bonds between water ligands and the central metal ion in hexaqua ions of metals such as $[Mg(H_2O)_6]^{2+}$ and $[Fe(H_2O)_6]^{2+}$
- Electrons do not 'circle around' nuclei — they exist as charge clouds or orbitals. Covalent bonds consist of overlapping orbitals. Ordinary covalent bonds are overlapping one-electron orbitals, whereas dative bonds are from the overlap of a two-electron orbital with an 'empty' orbital.

As a general rule the shorter the bond length the stronger is the covalent bond. Thus the C–Cl bond is stronger than the C–Br bond because chlorine has a smaller radius than bromine.

Covalent compounds that are small molecules have strong bonds *between atoms* — intramolecular bonds — but much weaker forces *between molecules*. They have, therefore, low melting and boiling temperatures because little energy is required to overcome the **intermolecular forces** (see page 22) between molecules.

Exam tip

Remember that anions (–) go to the anode (+), where they are oxidised by electron loss.

Exam tip

Covalent bonds are *not* broken when a molecular covalent substance, such as water, is melted or boiled.

Knowledge check 9

State the number of electrons in the outer orbit of a phosphorus atom in PCl_5.

Knowledge check 10

Draw how the 1s orbital of hydrogen overlaps with a 3*p* orbital of chlorine when forming a molecule of HCl.

Dot-and-cross diagrams

The electron structure of a covalent molecule can be drawn as a dot-and-cross diagram (Lewis structure), in which the electron pairs are shown as dots and crosses. The electrons are of course identical, so a diagram with all dots or all crosses would also be a correct representation. Some examples are shown in Figure 6.

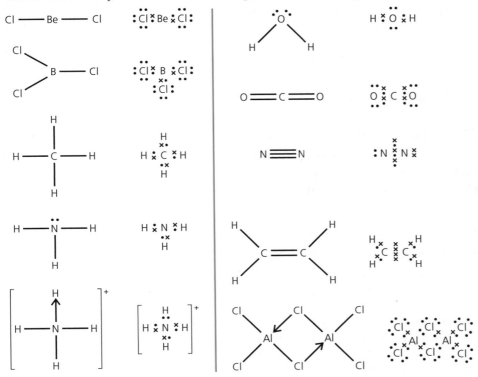

Figure 6

Metallic bonding

The Drude–Lorentz model of a metallic crystal shows the following features:

- Metal ions are usually held in a close-packed lattice, the exception being group 1 metals, which are not close-packed.
- The outer (valence) electrons are delocalised throughout the metal crystal.
- The bonding is not particularly directional. This means the lattice can be distorted without breaking, so the metal is malleable and ductile.
- The delocalised electrons are mobile and enable conduction of electricity in the solid.
- The more delocalised electrons there are, the stronger the bonding and the higher the melting temperature.
- The smaller the metal ion, the closer the packing and the higher the melting temperature.

Exam tip

A single line represents the shared pair of electrons in a single covalent bond.

Knowledge check 11

Draw a dot-and-cross diagram of PCl_3, showing outer electrons only.

Exam tip

Metals conduct electricity by movement of *electrons*, whereas ionic compounds conduct, but only when molten, by movement of *ions*.

Knowledge check 12

Explain why magnesium has a higher melting temperature than sodium.

Summary

After studying this part of the topic, you should be able to:
- explain an ionic bond, a covalent bond, a dative covalent bond and a metallic bond
- give evidence for the existence of ions
- explain why ionic compounds have high melting temperatures
- draw dot-and-cross diagrams for covalent molecules
- explain why metals conduct electricity and why potassium has a lower melting temperature than calcium

Shapes of molecules and ions

Valence-shell electron-pair repulsion theory

The shape of a molecule is determined by the repulsion of electron pairs around the central atom. There are several factors that have to be taken into account:
- The number of electron pairs around the atom of interest, both lone pairs and bond pairs.

Exam tip

The number of lone pairs can be worked out using the formula:

number of lone pairs = ½ × (group number of the element − number of covalent bonds)

- Lone pair–lone pair repulsions are greater than lone pair–bond pair repulsions, which in turn are greater than bond pair–bond pair repulsions. The repulsions between electrons modify the bond angles. Thus in methane the H–C–H angle is 109.5°, in ammonia the H–N–H angle is 107° and in water the H–O–H angle is 104°.

The shapes of analogous molecules can thus be predicted. The shape of methane (CH_4) is the same as that of silane (SiH_4). Phosphine (PH_3) has a similar shape to ammonia (NH_3), and the shape of water (H_2O) is similar to that of hydrogen sulfide (H_2S).

A double or triple bond is stereochemically equivalent to a single bond.

Species with single bonds only

The arrangement of the bonds depends on the number of electron pairs in total. The name of the shape depends on the position of the atom centres only (see Table 5).

Knowledge check 13

How many lone pairs are around the chlorine atom in the ClF_3 molecule?

Name	Formula	Bond pairs	Lone pairs	Shape	Structure
Beryllium chloride	$BeCl_2$	2	0	Linear	$Cl-Be-Cl$
Boron trichloride	BCl_3	3	0	Trigonal planar	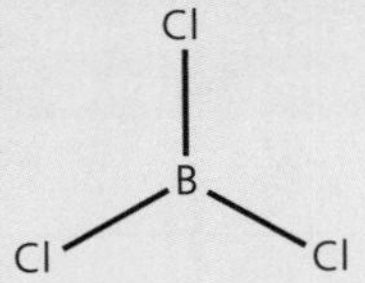
Methane	CH_4	4	0	Tetrahedral	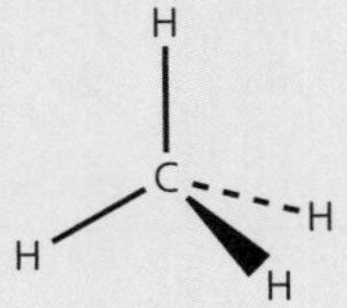
Hydrogen chloride	HCl	1	3	Linear	$H-Cl$
Ammonia	NH_3	3	1	Pyramidal	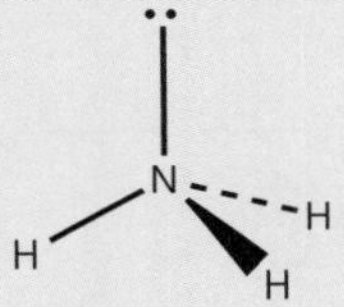
Ammonium ion	NH_4^+	4	0	Tetrahedral	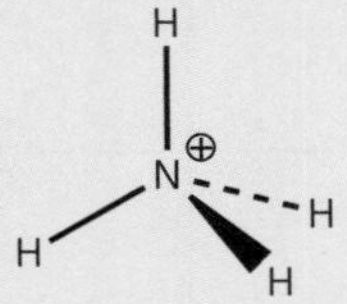
Water	H_2O	2	2	Bent	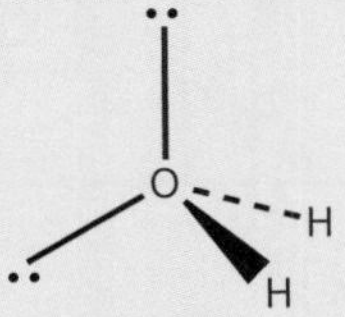
Phosphorus pentachloride (gas phase only — the solid is PCl_4^+ and PCl_6^- ions)	PCl_5	5	0	Trigonal bipyramidal	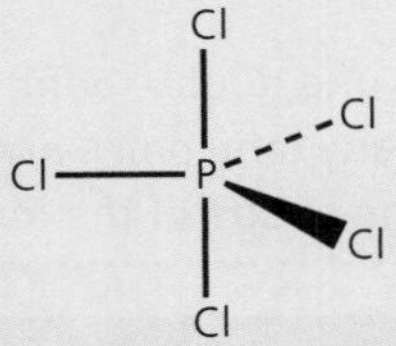
Sulfur hexafluoride	SF_6	6	0	Octahedral	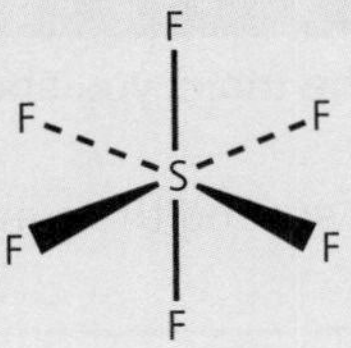

Table 5

Knowledge check 14

State the shape of the PCl_3 molecule.

Knowledge check 15

Which molecule has the smallest bond angle, PH_3, OF_2 or SiH_4?

Content Guidance

Species with double bonds

Stereochemically a double bond is equivalent to a single bond. Table 6 shows the structures of some species with double bonds.

Name	Formula	Single bonds	Double bonds	Lone pairs	Shape	Structure
Carbon dioxide	CO_2	0	2	0	Linear	$O=C=O$
Sulfur dioxide	SO_2	0	2	1	Bent	
Carbonate ion	CO_3^{2-}	2	1	0	Trigonal planar	
Sulfate ion	SO_4^{2-}	2	2	0	Tetrahedral	
Nitrate ion	NO_3^-	1	2	0	Trigonal planar	

Table 6

> ### Knowledge check 16
>
> The sulfite, SO_3^{2-}, ion has one double bond and two single bonds around the sulfur. How many lone pairs are there around the S atom and hence what is the shape of the ion?

Summary

After studying this part of the topic, you should be able to work out:

■ the number of lone pairs around the central atom in a molecule

■ the shape of the molecule
■ the bond angle

Intermediate bonding and bond polarity

Electronegativity and polarity

Ionic and covalent bonds are extremes. Ionic bonding involves complete electron transfer and covalent bonding involves equal sharing of electrons in pairs. In practice, bonding is intermediate between these two forms in most compounds, with one type being predominant.

Electronegativity

This is defined as the attraction an atom has for the bonding electrons in a covalent bond.

Fluorine is the most electronegative element, with a value of 4.0, caesium the least at 0.7. There are no units.

- Atoms with the same electronegativity bond covalently, with equal sharing of electrons.
- Atoms with different electronegativity form:
  - polar covalent bonds if the difference is not too large — up to about 1.5
  - ionic bonds if the difference is more than about 1.5

Polarity

Ionic compounds

Cations can distort the electron clouds of anions (which are generally larger). This distortion leads to a degree of electron sharing and hence some covalence in most ionic compounds.

- Small cations with a high charge (2+ or 3+) have a high charge density. They can polarise anions and have therefore a high polarising power.
- Large anions do not hold onto their outer electrons very tightly, so they can be easily distorted. Such ions are polarisable.
- Magnesium chloride ($MgCl_2$) has a polarising cation but a not very polarisable anion. There is some covalent character in the compound, but much more in magnesium iodide (MgI_2) since the large I^- ion is highly polarisable.

Polarity in molecules

A molecule may have polar covalent bonds, but will not be polar overall if the bond polarities cancel because of the shape of the molecule. Thus boron trifluoride (BF_3) has polar bonds but no overall polarity because the trigonal planar shape means that the bond polarities cancel (see Figure 7). The same is true for the tetrahedral carbon tetrachloride (CCl_4). Ammonia (NH_3) has polar bonds and is polar overall because its pyramidal shape does not cause the individual polarities to cancel.

Exam tip

Do not say that the charges cancel. It is the polarities that cancel.

These molecules have polar bonds, but the polarities cancel overall because of the symmetry of the molecule.

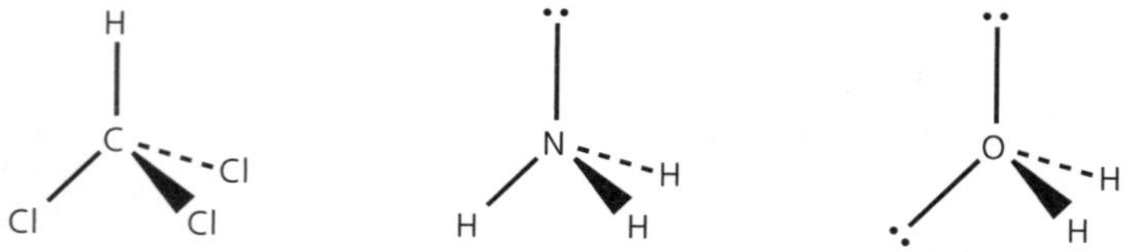

These molecules also have polar bonds, but the individual polarities do not cancel so the molecules are polar overall.

Figure 7

Knowledge check 17

Use the data booklet to find the electronegativities and then state which of the following molecules contain polar bonds: CH_4, CO_2, OF_2, CH_2Cl_2, CI_4 (carbon tetraiodide).

Knowledge check 18

Which of the molecules in Knowledge check 17 are polar?

Summary

After studying this part of the topic, you should be able to:
- define electronegativity
- suggest whether a substance is mostly ionic or polar covalent
- explain whether a molecule is polar or not

Intermolecular forces

Types of intermolecular force

There are several types of intermolecular force. They depend to some extent on the distortion of the electron distribution of one molecule by another. At close range there are also repulsive forces in action. In decreasing order of strength, the attractive forces are:
- hydrogen bonds
- instantaneous dipole–induced dipole forces (usually called London or dispersion forces)
- permanent dipole–permanent dipole forces

Hydrogen bonds

Hydrogen bonds are electrostatic forces between a hydrogen atom covalently bonded to a small electronegative atom (N, O or F) and another N, O or F atom, usually on a different molecule. As the hydrogen is highly $\delta+$, it is then attracted to *a lone pair of electrons* on the electronegative atom, which is $\delta-$. The bond angle of the three atoms involved is 180°.

Hydrogen bond strength in hydrogen fluoride (HF) is around $150\,kJ\,mol^{-1}$, but most others are between $60\,kJ\,mol^{-1}$ and $20\,kJ\,mol^{-1}$. Hydrogen bonding is responsible for the high boiling temperatures of NH_3, H_2O and HF compared with the other hydrides in their groups and for the high solubility of alcohols and sugars in water. The hydrogen bonds formed between the solute and the water are of similar strength to the hydrogen bonds in water itself.

Ice forms crystals having a hexagonal lattice structure. Each water molecule is surrounded by four neighbouring water molecules. Two of these are hydrogen-bonded to the oxygen atom on the central H_2O molecule, and each of the two hydrogen atoms is similarly bonded to another neighbouring H_2O. The distance across the hexagon is greater than between neighbouring molecules. The result is that ice is less dense than liquid water.

Instantaneous dipole–induced dipole forces

These are present between *all* molecules and are usually called **London** forces. They are the only interaction in non-polar molecules or in single atoms such as the monatomic inert or noble gases. These forces arise from a temporary dipole inducing a complementary dipole in an adjacent molecule. These dipoles are always shifting, but are induced in phase and give a net attraction. Their strength depends on the number of electrons in the molecule and its shape, which determine the polarisability of the molecule.

Permanent dipole–permanent dipole forces

These are normally *weaker* than London forces. The polar molecules attract via their permanent dipoles. Polar molecules have higher boiling temperatures than non-polar molecules with a similar number of electrons.

The term **van der Waals forces** is sometimes used to describe the total of London and permanent dipole interactions.

Properties of materials

Intermolecular and inter-ion forces (be clear which you are talking about) determine the physical properties of a substance such as its melting and boiling temperatures, hardness and density.

Change of state

In solids the particles vibrate about a mean position in the crystal lattice. Heating increases the amplitude of the vibration until a point where the interparticular forces are overcome and the lattice collapses — the solid melts. Just above the melting temperature, liquids often have considerable order to their structure; further heating causes movement over increasing distances, until the temperature is such that the vapour pressure of the liquid is the same as the external pressure. At this temperature the interparticular forces are overcome and the liquid boils.

Noble gases

The only interatomic forces are London forces arising from the limited movement of electrons within the atom. The smaller the atom, the less the movement and so the weaker the attractions. This is reflected in their boiling temperatures (Table 7).

Element	He	Ne	Ar	Kr	Xe	Rn
Boiling temperature/K	4	27	87	121	165	211

Table 7 Boiling temperatures of the noble gases

Exam tip

Hydrogen bonds occur between molecules containing O–H, N–H or H–F bonds.

Exam tip

For molecules that do *not* hydrogen bond but have similar numbers of electrons, the relative boiling temperatures are determined by dipole forces. If they have a very different number of electrons, the London forces determine the relative boiling points.

Knowledge check 19

List all the intermolecular forces present between HCl molecules.

Hydrides of group 4

- None of the elements in group 4 is electronegative enough to give hydrogen bonding.
- The hydrides have strong covalent intramolecular bonds and weak London intermolecular forces.
- The more electrons, the greater the attraction. The boiling temperatures are shown in Table 8.

Hydride	Methane (CH_4)	Silane (SiH_4)	Germane (GeH_4)	Stannane (SnH_4)
Boiling temperature/K	109	161	183	221

Table 8 Boiling temperatures of group 4 hydrides

Hydrides of groups 5–7

Nitrogen, oxygen and fluorine are electronegative enough and small enough to give hydrogen bonding in their hydrides. Therefore, for molecules of their size, these hydrides have much higher boiling temperatures than would be expected from van der Waals interactions alone. The remaining hydrides have both dipole–dipole and London interactions and in every case the latter dominate. Thus their boiling temperatures increase with increasing number of electrons in the hydride (Table 9).

Group 5 hydride	NH_3	PH_3	AsH_3	SbH_3
Boiling temperature/K	240	185	218	256

Group 6 hydride	H_2O	H_2S	H_2Se	H_2Te
Boiling temperature/K	373	212	232	269

Group 7 hydride	HF	HCl	HBr	HI
Boiling temperature/K	293	188	206	238

Table 9

> ### Knowledge check 20
>
> Which force is more important in determining the boiling temperatures of the hydrides PH_3 to SbH_3?

> ### Exam tip
>
> Although the chlorine atom is very electronegative, it is too big for any significant amount of hydrogen bonding.

Alkanes

The intermolecular forces in the non-polar alkanes are London forces. They increase in strength with increasing numbers of electrons in the molecules. There is an increase in both the melting and the boiling temperatures of the straight-chain alkanes as the chain length, and hence the number of electrons, increases. The values for the melting temperature of the solids are less regular at first because of differences in the crystal packing. The values in Table 10 are given to three significant figures.

Alkane	Methane	Ethane	Propane	Butane	Pentane	Hexane
Boiling temperature/K	109	185	231	273	309	342

Table 10

Effect of branching

Since the strength of the intermolecular forces falls off rapidly with increasing distance, the boiling temperatures of branched-chain isomers of the alkanes are generally lower than those of the straight-chain compound (see below). The London forces are weaker with branched chain isomers, because they cannot pack as well and so have fewer points of contact. Table 11 compares the boiling temperatures of the three isomers of C_5H_{12}.

Alkane	Pentane	2-methylbutane	2,2-dimethylpropane
Boiling temperature/K	309	301	283

Table 11

Volatility of alkanes and alcohols

An alcohol is less volatile (i.e. has a higher boiling temperature) than an alkane with a similar number of electrons (Table 12). Alcohols are strongly hydrogen-bonded, whereas alkanes are non-polar and have only London forces between their molecules. These London forces are much weaker than hydrogen bonds.

Alkane	Ethane (CH_3CH_3) (18 electrons)	Propane ($CH_3CH_2CH_3$) (26 electrons)	Butane ($CH_3CH_2CH_2CH_3$) (34 electrons)
Boiling temperature/K	185	231	273
Alcohol	Methanol (CH_3OH) (18 electrons)	Ethanol (CH_3CH_2OH) (26 electrons)	Propan-1-ol ($CH_3CH_2CH_2OH$) (34 electrons)
Boiling temperature/K	338	352	371

Table 12

Solute, solvent and solubility

The solubility of a substance in a given solvent depends on the enthalpy (heat energy) and entropy changes of the dissolving process. Entropy is considered in the second year of the A level course, so solubility is discussed here only in terms of enthalpy, but you should remember that the full story is more complex.

In general a substance dissolves if the forces between the solute and the solvent are of similar or greater strength than those that are broken between both the solute molecules and the solvent molecules. Thus, for a solute A and a solvent B, the A–B forces must be similar in strength to, or stronger than, both the A–A forces and the B–B forces.

Ionic compounds

When an ionic compound dissolves in water, the ionic lattice has to be broken up (endothermic process) and the resulting ions are then hydrated (exothermic process). If the energy required to break up the lattice is recovered by the hydration of the ions, the compound is soluble.

Exam tip

When comparing volatility or boiling temperatures, see if one compound forms hydrogen bonds. Then look for the one with the largest number of electrons and, lastly, see if one is polar.

Knowledge check 21

Ethanol, C_2H_5OH, and dimethyl ether, CH_3OCH_3, are isomers. Explain which is more soluble in water.

Alcohols in water

An alcohol molecule (ROH) consists of a non-polar alkyl group and a hydroxyl group (–OH) that can hydrogen-bond with water. On dissolving, the –OH group breaks some of the water–water hydrogen bonds and replaces them with alcohol–water hydrogen bonds. The alcohol dissolves if the new alcohol–water bonds formed are strong enough to compensate energetically for the water–water bonds that are broken by the alkyl group (R) occupying space in the liquid water.

Alcohols up to C_3 are miscible with water in all proportions. Butan-1-ol and its isomers are partially soluble, dissolving to some extent; alcohols above C_5 are insoluble in water. The long hydrophobic carbon chains disrupt the structure of the water too much.

Larger alcohols dissolve if they have more –OH groups. For example, glucose is a C_6 molecule, but it has six –OH groups, so it is highly soluble in water.

Polar molecules in water

Hydrogen bonding in water is strong. To overcome this and separate the water molecules a solute has to form strong bonds with water. Polar molecules such as halogenoalkanes are not usually able to form strong enough bonds, so they are not water-soluble.

Non-aqueous solvents: 'like dissolves like'

Since the intermolecular forces in the solute are similar to those between the solvent and to those between solvent and solute in the solution, non-polar solvents such as hydrocarbons (e.g. hexane) dissolve other non-polar substances, such as waxes and fats. In contrast, hydrocarbons do not dissolve sugars, because the hydrogen bonding between the sugar molecules in the solid is too strong to be energetically compensated by any interaction between the non-polar hydrocarbon and the polar sugar molecules.

Exam tip

Hydrogen-bonded compounds, such as alcohols, will dissolve in water unless they have a long non-polar chain. Propanone and non-polar solvents will dissolve compounds where the main intermolecular forces are London forces.

Knowledge check 22

Explain why the polar propanone, CH_3COCH_3, which contains a $\delta-$ oxygen, is soluble in water.

Knowledge check 23

Which of the following is the most soluble in water and which is the most soluble in hexane: $C_2H_5NH_2$, C_2H_5Cl and C_2H_6?

Summary

After studying this part of the topic, you should be able to explain:
- hydrogen bonding and predict which compounds can form this type of intermolecular force
- the trend in boiling temperatures of the group 4, 5, 6 and 7 hydrides
- the difference in boiling temperatures of a series of alkanes
- the solubility of alcohols and ionic compounds in water

Topic 2B Structure

The structure of solids can be divided into four types:
- giant ionic lattices
- simple molecular
- giant covalent lattices
- metals

Giant ionic lattices

All solid ionic substances consist of a regular three-dimensional lattice of alternating positive and negative ions (see Figure 8).

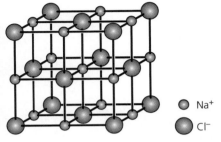

Na⁺

Cl⁻

Figure 8 The ionic structure of sodium chloride

Strength of ionic bonds

An ionic bond is the strong electrostatic force of attraction between oppositely charged ions.

An ionic crystal consists of a giant three-dimensional lattice of ions. The structure depends on the relative size of the anion and cation, and on the stoichiometry of the substance. Clearly, NaCl (Figure 8) cannot have the same crystal structure as $MgCl_2$.

The crystal is held together by a strong *net* attractive force between oppositely charged cations and anions. There are also longer-range repulsive forces between ions of the same charge. However, because these ions are more widely separated the repulsive forces are not as large as the attractive forces between ions of different charge.

The attractive force between two ions is described by Coulomb's law:

$$|F| = \frac{Q_1 Q_2}{k(r_+ + r_-)^2}$$

where $|F|$ is the magnitude of the force between the ions, Q_1 and Q_2 are the charges on the two ions, k is a constant, and $r_+ + r_-$ is the sum of the radii of the two ions and is, therefore, the distance between the centres of the ions when they are touching in the crystal. For a given crystal structure, the attractive force depends on the product of the two charges, and inversely on the square of the distance between the ion centres. This means that small and highly charged ions have much stronger forces of attraction and so a higher melting point.

Formation of an octet alone does *not* make an ion more stable than the atom from which it arose.

Simple molecular

These comprise elements and compounds consisting of simple molecules. Examples are ice, iodine and almost all solid organic substances. They have a regular arrangement of molecules, with strong covalent bonds within each molecule and weak intermolecular forces between different molecules. In order to melt the solid, the weak intermolecular forces, *not* the stronger covalent bonds, need to be broken, so they have low melting temperatures.

Knowledge check 24

Explain why ionic solids have high melting temperatures.

Giant covalent lattices

Diamond and graphite consist of a regular arrangement of carbon atoms covalently bonded together. To melt them, strong covalent bonds have to be broken.

- **Diamond** has layers of hexagonal rings that are puckered (i.e. not flat); each carbon atom is bonded covalently to four others throughout the lattice. A great deal of energy is needed to break these bonds, so diamond has an extremely high melting temperature (about 3800°C). There are no free electrons, so diamond is a poor electrical conductor. It is a good thermal conductor, since the rigid lattice readily transmits vibrations.
- **Graphite** has layers of flat hexagons with a fourth bond delocalised along the plane of the carbons. The forces between the layers are London forces and are, therefore, weak. Graphite is really a stack of giant molecules, rather like a pile of paper. It is a good electrical conductor parallel to the planes of carbon atoms, but a poor thermal conductor. It sublimes at 3730°C.
- **Graphene** is pure carbon in the form of a very thin, nearly transparent sheet, one atom thick. It is remarkably strong for its very low weight (100 times stronger than steel) and it conducts heat and electricity with great efficiency. While scientists had theorised about graphene for decades, it was first produced in the laboratory in 2004. It is two-dimensional and can be described as a one-atom-thick layer of graphite.
- **Buckminsterfullerene** (C_{60}) was discovered in the late 1990s. It is a sphere of pentagons and hexagons, exactly like a football. It is red and soluble in organic solvents.
- **Carbon nanotubes** are related to fullerenes. They have tubular structures, made from carbon hexagons, with diameters from one to several nanometres. Their structure is like a roll of wire netting. They can be capped at each end with a hemisphere of carbon atoms to make nanometre-sized capsules. Much research is being conducted to develop these tiny capsules so that they can deliver drugs and other treatments to tumour cells with great precision, avoiding damage to healthy tissue. Another potential application is in building electronic devices even smaller than those presently available.

Metals

These are a giant structure of metal ions, usually face- or body-centred cubic, or hexagonal close-packed. Arranged randomly between the metal ions are the delocalised electrons. Sodium has one delocalised electron per ion and calcium two.

The electrostatic force of attraction between the metal ions and the delocalised electrons varies from weak in sodium to strong in iron. The strength depends on:

- the crystal structure
- the radius of the metal ion (the metallic radius)
- the charge on the metal ion (this equals the number of delocalised electrons per atom)

Physical properties and structure

Table 13 shows the relationship between physical properties and structure.

Property	Structural type
Solid with a low melting temperature and soluble in water	Simple molecular Forms hydrogen bonds with water
Solid with a low melting temperature and insoluble in water	Simple molecular Cannot form hydrogen bonds with water
Extremely high melting temperature and insoluble in all solvents	Giant covalent
High melting temperature Conducts electricity when molten but *not* when solid May be water soluble	Ionic
Variety of melting temperature Conducts electricity when solid (and when molten) Only dissolves in other metals (alloys)	Metallic

Table 13

Summary

After studying this part of the topic, you should be able to:

- identify the different types of giant structure and the physical properties associated with them
- give examples of simple molecular structures

- describe the different structures of carbon
- predict melting and boiling temperatures, electrical conductivity, solubility and types of intermolecular force associated with types of bonding and structure

■ Topic 3 Redox I

Oxidation numbers

Oxidation numbers are used to find the ratio (the stoichiometry) in which the oxidising and reducing agents react. To find the oxidation number of an element in any species, the following rules are used:

- An uncombined element has an oxidation number of zero.
- A simple ion has an oxidation number equal to its charge.
- Hydrogen has oxidation number +1, except in metal hydrides where it is −1.
- Oxygen has oxidation number −2, except in peroxides, where it is −1, and when bonded to the more electronegative fluorine, where it is +2.
- The *sum* of the oxidation numbers in a neutral molecule is zero.
- The *sum* of the oxidation numbers in a multi-atom ion equals the charge on that ion.

Exam tip

Peroxides contain an O–O bond, so the oxygen atoms in Na_2O_2 and H_2O_2 are −1 each.

Example 1

What is the oxidation number of sulfur in sulfur trioxide (SO_3)?

Answer

Oxygen has an oxidation number of -2, which gives $S^{x+}(O^{2-})_3$, so $x = +6$. The oxidation number of S in SO_3 is $+6$.

Example 2

What is the oxidation state of manganese in manganate (MnO_4^-)?

Answer

A similar argument gives $[Mn^{y+}(O^{2-})_4]^-$. The oxygen contributes -8, with -1 charge overall. Thus $y = +7$. The oxidation number of Mn in MnO_4^- is $+7$.

The oxidation number of an atom in a compound is the charge that it would have if the compound were ionic. It is useful because changes in oxidation number indicate that an atom has been oxidised or reduced. Oxidation numbers can be used for simple atoms and ions, molecules or complex ions.

The addition of a Roman numeral in brackets after the name of an element indicates its oxidation number. Thus $FeCl_3$ is iron(III) chloride and MnO_4^- ions are manganate(VII) ions.

Electron transfer

The reaction of magnesium with oxygen is obviously an oxidation:

$$2Mg + O_2 \rightarrow 2MgO$$

- Magnesium is converted into magnesium ions Mg^{2+}, the oxygen to oxide ions O^{2-}.
- A logical extension of this reaction is to say that any reaction converting magnesium to its ion is an oxidation and that **oxidation is electron loss**.
- The reverse process, the gain of electrons, is therefore **reduction**.
- Remember **OILRIG**: **o**xidation **i**s **l**oss of electrons, **r**eduction **i**s **g**ain of electrons.

An oxidising agent removes electrons from another species, so it gains electrons in a redox process and is itself reduced. Similarly, a reducing agent is itself oxidised and therefore loses electrons.

Changes in oxidation number

Oxidation occurs if the oxidation state of an atom rises; reduction causes the oxidation state to fall.

Knowledge check 25

What are the oxidation numbers of chromium in the $Cr_2O_7^{2-}$ ion and nitrogen in the NH_4^+ ion?

Exam tip

Metals form positive ions as they become oxidised by electron loss and when non-metals form negative ions they are reduced by electron gain.

Exam tip

An oxidising agent gets reduced and a reducing agent gets oxidised.

Example

Iron reacts with chlorine on heating to give iron(III) chloride.

$$2Fe \quad + \quad 3Cl_2 \quad \rightarrow \quad 2FeCl_3$$

Oxidation numbers: 0   0   +3 −1

The oxidation number of the iron goes up, so it has been oxidised.
The oxidation number of chlorine goes down, so it has been reduced.

Disproportionation

Disproportionation is a reaction where the oxidation number of an element in a *single* species both rises and falls.

Example

In the disproportionation of chlorate(I) ions (OCl^-), the oxidation number of chlorine both rises and falls:

$$3OCl^- \rightarrow \quad 2Cl^- \quad + \quad ClO_3^-$$

Oxidation number of chlorine   +1   −1   +5

Two chlorine atoms in the +1 state are reduced to the −1 state and one is oxidised from +1 to +5.

Combining half-equations

Redox reactions are a combination of an oxidation and a reduction process. Each process can be written as a half-equation, which includes electrons. Two half-equations can be combined to give an overall equation. One or both of the half-equations may need to be multiplied by a small integer in order to make the number of electrons the same in both equations. This is because *the same electrons* are involved in each equation.

Example 1

The reaction of chlorine with the bromide ions in seawater is used in the commercial production of bromine.

Oxidation of bromide ions: $2Br^- \rightarrow Br_2 + 2e^-$

Reduction of chlorine: $Cl_2 + 2e^- \rightarrow 2Cl^-$

Since there are two electrons in both half-reactions, they can simply be added to give the overall equation:

$$2Br^- + Cl_2 \rightarrow Br_2 + 2Cl^-$$

Knowledge check 26

Explain, in terms of oxidation numbers, what has been oxidised and what reduced in the equation:

$SnCl_2 + 2FeCl_3 \rightarrow$
$SnCl_4 + 2FeCl_2$

Exam tip

The total increase of oxidation number of the element being oxidised must equal the total decrease of the element reduced.

Example 2

Iron reacts with chlorine on heating to give iron(III) chloride:

Oxidation of iron: $Fe \rightarrow Fe^{3+} + 3e^-$

Reduction of chlorine: $Cl_2 + 2e^- \rightarrow 2Cl^-$

Since the number of electrons is not the same, the top reaction is multiplied by 2 and the bottom one by 3. The equations can then be added to give the overall reaction:

$2Fe \rightarrow 2Fe^{3+} + 6e^-$

$3Cl_2 + 6e^- \rightarrow 6Cl^-$

Overall:

$2Fe + 3Cl_2 \rightarrow 2Fe^{3+} + 6Cl^-$ (or $2FeCl_3$)

Exam tip

Make sure that you cancel the electrons when writing the overall equation.

Summary

After studying this topic, you should be able to:
- work out oxidation numbers given the formula of a molecule or ion
- explain oxidation in terms of electron loss or change of oxidation number
- define disproportionation
- combine redox half-equations to form the overall balanced equation

■ Topic 4 Inorganic chemistry and the periodic table

Topic 4A Group 2 (alkaline earth metals)

Ionisation energies

The ionisation energies fall with increasing atomic number (down the group). Although the nuclear charge is rising, the size of the atom is also rising, as is the amount of shielding (or repulsion) from inner-shell electrons. Since the outer electrons are further from the nucleus they become easier to remove.

Because metals react by loss of electrons (forming positive ions), the metals in group 2 become more reactive down the group as their ionisation energy decreases (Table 14).

Element	Be	Mg	Ca	Sr	Ba
First ionisation energy/kJ mol⁻¹	900	736	590	548	502
Second ionisation energy/kJ mol⁻¹	1760	1450	1150	1060	966

Table 14

Reactions of group 2 metals

Reaction with oxygen

All the group 2 metals burn to produce the oxide. For example:

$$2Mg(s) + O_2(g) \xrightarrow{burn} 2MgO(s)$$

- Magnesium burns vigorously with a brilliant white flame.
- Calcium burns with a brick-red flame.
- Strontium burns with a crimson flame.
- Barium burns with a pale apple-green flame, forming substantial amounts of barium peroxide (Ba_2O_2), as well as the oxide (BaO).

All the group 2 oxides are white, ionic compounds.

Reaction with chlorine

On heating with chlorine, all the group 2 metals react similarly to give white, ionic chlorides:

$$Mg(s) + Cl_2(g) \xrightarrow{heat} MgCl_2(s)$$

Reaction with water

The vigour of the reaction decreases down the group.

- Magnesium reacts slowly with cold water, but rapidly with steam.

$$Mg(s) + H_2O(g) \xrightarrow{heat} MgO(s) + H_2(g)$$

- Calcium reacts quite quickly with cold water to give a milky suspension of calcium hydroxide, some of which dissolves.

$$Ca(s) + 2H_2O(l) \rightarrow Ca(OH)_2(aq) + H_2(g)$$

- Strontium and barium react similarly, the reaction of barium being vigorous and giving a colourless solution of barium hydroxide — the most soluble of the group 2 hydroxides.

Reaction of group 2 oxides with water

The oxides of group 2 metals react with water to give hydroxides. For example:

$$MgO(s) + H_2O(l) \rightarrow Mg(OH)_2(aq \text{ and } s)$$

Magnesium hydroxide is sparingly soluble. The solubility of the hydroxides increases down group 2.

Reactions of group 2 oxides and hydroxides with dilute acids

All the group 2 oxides are basic. The reaction of magnesium oxide is typical:

$$MgO(s) + 2H^+(aq) \rightarrow Mg^{2+}(aq) + H_2O(l)$$

Reaction of the other group 2 oxides with sulfuric acid produces a coating of the insoluble metal sulfate, which prevents further reaction.

> **Exam tip**
>
> Group 2 metals form 2+ ions in all their compounds.

> **Exam tip**
>
> Group 2 oxides are strong enough bases to deprotonate water and form a hydroxide.

The basic hydroxides react similarly with acid:

$$Mg(OH)_2(s) + 2H^+(aq) \rightarrow Mg^{2+}(aq) + 2H_2O(l)$$

Solubility of group 2 hydroxides and sulfates

You need only know the trends in solubility of the group 2 hydroxides and sulfates — you are not required to explain them. The solubility values are given in moles of solute per dm^3 water at 25°C.

Hydroxides

The solubility of the hydroxides *increases* with increasing atomic number of the cation (Table 15).

Compound	$Mg(OH)_2$	$Ca(OH)_2$	$Sr(OH)_2$	$Ba(OH)_2$
Solubility/mol dm^{-3}	2.00×10^{-5}	1.53×10^{-3}	3.37×10^{-3}	1.50×10^{-2}

Table 15

Sulfates

The solubility of the sulfates *decreases* with increasing atomic number of the cation (Table 16). In practice, only magnesium sulfate is noticeably soluble.

Compound	$MgSO_4$	$CaSO_4$	$SrSO_4$	$BaSO_4$
Solubility/mol dm^{-3}	1.83×10^{-1}	4.66×10^{-3}	7.11×10^{-5}	9.43×10^{-7}

Table 16

Thermal stability

The ease of decomposition of an ionic nitrate or carbonate depends on the polarising power of the metal cation. The smaller the ionic radius and the greater the charge, the stronger is the polarising power of the cation. This results in a lower decomposition temperature compared with a cation with a smaller polarising power. Thus stability increases down a group as the ions get larger and group 1 compounds are more stable to heat than equivalent group 1 compounds in the same period.

Nitrates of groups 1 and 2

Nitrates decompose on heating to give either the metal nitrite and oxygen (sodium to caesium in group 1):

$$2NaNO_3 \rightarrow 2NaNO_2 + O_2$$

or nitrogen dioxide, oxygen and the metal oxide (lithium in group 1, magnesium to barium in group 2):

$$2LiNO_3 \rightarrow Li_2O + 2NO_2 + \tfrac{1}{2}O_2$$

$$M(NO_3)_2 \rightarrow MO + 2NO_2 + \tfrac{1}{2}O_2$$

where M is a group 2 metal.

The ease of decomposition is related to the polarising power of the cation. Thus the small, doubly charged cation, Mg^{2+}, in magnesium compounds polarises the anion and makes decomposition easier. The larger Ba^{2+} cation causes less polarisation and the decomposition is more difficult than with other group 2 compounds. Group 1 nitrates decompose with difficulty to the nitrite, except lithium nitrate, which has the smallest and therefore the most polarising cation in this group (Li^+).

Carbonates of groups 1 and 2

Group 2 carbonates decompose on heating to give the metal oxide and carbon dioxide. Decomposition becomes more difficult down the group, as the cation gets bigger and less polarising. The magnesium ion (Mg^{2+}) is the smallest and therefore the most polarising cation in group 2, so magnesium carbonate is thermally the least stable.

Group 1 carbonates, except lithium carbonate, do not decompose when heated.

This trend can be demonstrated by placing some of a group 2 carbonate in a test tube fitted with a delivery tube dipping into limewater. The test tube is heated and the time for the limewater to go milky is noted. The experiment is repeated with the same amount of a different carbonate, heating with the same flame and measuring the time for the limewater to go milky.

The decomposition of calcium carbonate is important in the blast furnace for the production of iron and other metals and in the manufacture of Portland cement.

$$CaCO_3(s) \rightarrow CaO(s) + CO_2(g)$$

In a closed system this reaction is in equilibrium.

Flame tests

If the atoms or ions of some group 1 or 2 elements are heated in a Bunsen flame, they emit light of distinctive colours (Table 17). The flame excites electrons to higher energy orbitals. When the electrons fall back down, they emit light of characteristic wavelength.

Metal	Lithium	Sodium	Potassium	Calcium	Strontium	Barium
Colour	Crimson red	Yellow	Lilac	Orange-red or brick red	Carmine red	Apple green

Table 17 Flame test colours of some group 1 and 2 elements

The flame colours can be used quantitatively to determine ion concentrations (e.g. in blood or urine) using a flame photometer.

The technique for a flame test is as follows:

- Dip a platinum wire in concentrated hydrochloric acid and place in the hottest part of a Bunsen flame. If the flame is coloured, repeat until the wire does not colour the flame.
- Dip the wire in concentrated hydrochloric acid and then into the solid to be tested. Place in the hottest part of the flame and observe the colour.

Exam tip

The same technique can be used to demonstrate the ease of decomposition of nitrates, with the time needed for brown fumes to be observed noted.

Knowledge check 27

Write the equation for the action of heat on lithium carbonate.

Exam tip

Why the flame is coloured is frequently asked.

Knowledge check 28

A white solid gave an orange-red flame colour and when heated decomposed to form a single acidic gas. Name the compound.

Summary

After studying this topic on group 2 you should be able to:

- write equations for the reactions of the metals with oxygen, water and the halogens and for the reactions of their oxides with water and acids

- state the trends in the solubility of the hydroxides and sulfates
- state and explain the trends in thermal stability of the carbonates and nitrates
- describe the flame colours and explain how the colour occurs

Topic 4B Group 7 (halogens)

Physical properties of the elements

The physical properties of the group 7 elements are summarised in Table 18.

Element	State at room temperature	Melting temperature/K	Boiling temperature/K
Fluorine	Pale yellow gas	53.5	85.0
Chlorine	Greenish-yellow gas	172	239
Bromine	Brown volatile liquid	266	332
Iodine	Dark grey lustrous solid; the vapour is purple	387	457

Table 18

The melting temperature of the halogens increases from F_2 to I_2. This is because the number of electrons in the molecules increases, causing the London intermolecular forces to get stronger. Thus more energy is required to separate the molecules.

Iodine sublimes when heated with a Bunsen burner as it causes such a rapid temperature rise that the liquid phase is not usually seen.

Chlorine and bromine dissolved in water or in organic solvents give pale green and orange or yellow solutions respectively. Iodine is purple when dissolved in liquids that do not contain oxygen in the molecule (e.g. hexane or benzene). In aqueous or alcoholic solution, iodine is brown. It is not very soluble in water. Addition of potassium iodide to aqueous iodine gives a brown solution of potassium triiodide (KI_3), often (but inaccurately) labelled 'iodine solution':

$$I_2(aq) + KI(aq) \rightleftharpoons KI_3(aq)$$

Iodine also turns a solution of starch an intense blue-black colour.

Oxidising reactions of the halogens

The electronegativity of the halogens decreases from fluorine (the most electronegative of all the elements in the periodic table) to iodine. As they react by gain of electrons (going from the zero oxidation state to the −1 state), their reactivity *decreases* down the group.

Exam tip

Turning starch blue-black is a test for iodine.

Halogens are oxidising agents. Their oxidising power decreases in the order chlorine > bromine > iodine. Oxidising agents are reduced (gain electrons) when they react, so the smaller the halogen atom, the more strongly attracted the electron gained will be.

Reactions with metals

On heating, metals react with halogens to form halides. The following reactions are typical of the halogens. Unless otherwise shown, the reactions with bromine and iodine are the same as those with chlorine.

$$2Na(l) + Cl_2(g) \rightarrow 2NaCl(s)$$

$$Ca(s) + Cl_2(g) \rightarrow CaCl_2(s)$$

Iron can form Fe^{2+} or Fe^{3+} ions. The reaction of iron with chlorine gives iron(III) chloride:

$$2Fe(s) + 3Cl_2(g) \rightarrow 2FeCl_3(s)$$

Reaction of iron with the less powerful oxidising agent iodine gives iron(II) iodide:

$$Fe(s) + I_2(g) \rightarrow FeI_2(s)$$

Reactions with non-metals

Some non-metals react with halogens and give covalent compounds. Hydrogen burns in an atmosphere of chlorine or bromine. Mixtures of the gases explode if heated.

$$H_2(g) + Cl_2(g) \rightarrow 2HCl(g)$$

White phosphorus, on heating with chlorine, initially gives phosphorus trichloride. This product reacts with excess chlorine to form phosphorus pentachloride:

$$P_4(s) + 6Cl_2(g) \rightarrow 4PCl_3(l)$$

$$PCl_3(l) + Cl_2(g) \rightleftharpoons PCl_5(g)$$

Reactions of aqueous halogens with reducing agents

In aqueous solution, chlorine oxidises iron(II) ions to iron(III) ions:

$$2Fe^{2+}(aq) + Cl_2(aq) \rightarrow 2Fe^{3+}(aq) + 2Cl^-(aq)$$

Bromine reacts similarly but more slowly.

Oxidation of halide ions by halogens: displacement reactions

Chlorine oxidises (displaces) both bromide and iodide ions. Bromine, a less powerful oxidising agent than chlorine, oxidises iodide ions. The first reaction is used to manufacture bromine from seawater:

$$Cl_2(aq) + 2Br^-(aq) \rightarrow 2Cl^-(aq) + Br_2(aq)$$

$$Cl_2(aq) + 2I^-(aq) \rightarrow 2Cl^-(aq) + I_2(aq)$$

$$Br_2(aq) + 2I^-(aq) \rightarrow 2Br^-(aq) + I_2(aq)$$

Exam tip

The group 1 elements are oxidised from the zero state to the +1 state and the group 2 elements to the +2 state.

Exam tip

The equation $2P + 3Cl_2 \rightarrow 2PCl_3$ will be accepted.

Exam tip

Iodine does not oxidise Fe^{2+} ions. Instead, iodide ions are oxidised by Fe^{3+} ions to iodine.

Disproportionation reactions

Disproportionation is the simultaneous oxidation and reduction of an element in a *single* species. The halogens form oxyanions in which the halogen has a positive oxidation state. Thus chlorate(I) ions (OCl^-) contain chlorine in the +1 state and chlorate(V) ions (ClO_3^-) have chlorine in the +5 state.

Reaction of chlorine with water or sodium hydroxide solution causes it to disproportionate.

- chlorine with water:

$$Cl_2(aq) + H_2O(l) \rightarrow HOCl(aq) + HCl(aq)$$
$$0 \qquad\qquad\qquad +1 \qquad -1$$

This reaction is useful for purifying drinking water. Both the dissolved chlorine and the chloric(I) acid, HOCl, will oxidise harmful bacteria.

- chlorine with cold dilute NaOH solution:

$$Cl_2(aq) + 2NaOH(aq) \rightarrow NaCl(aq) + NaOCl(aq) + H_2O(l)$$
$$0 \qquad\qquad\qquad\qquad -1 \qquad\qquad +1$$

- chlorine with hot concentrated NaOH solution:

$$3Cl_2(aq) + 6NaOH(aq) \rightarrow 5NaCl(aq) + NaClO_3(aq) + 3H_2O(l)$$
$$0 \qquad\qquad\qquad\qquad -1 \qquad\qquad +5$$

Chlorate(I) ions disproportionate on heating in solution to give chlorate(V) and chloride:

$$3OCl^-(aq) \rightarrow ClO_3^-(aq) + 2Cl^-(aq)$$

The halogens, except fluorine, behave identically in these reactions.

Reactions of halide salts

Reaction with concentrated sulfuric acid

Sulfuric acid is a stronger acid than the hydrogen halides (HX, where X = Cl, Br or I). The hydrogen halides are also gases, so if concentrated sulfuric acid is added to a potassium halide, HX is liberated, since sulfuric acid donates a hydrogen ion to the halide ion. For example, with KCl:

$$KCl(s) + H_2SO_4(l) \rightarrow KHSO_4(s) + HCl(g)$$

Steamy fumes of HCl are given off.

In the case of bromide and iodide salts, the HBr or HI liberated is a strong enough reducing agent to be oxidised by sulfuric acid, so little hydrogen halide is obtained. Bromides are oxidised to bromine and the sulfuric acid is reduced to sulfur dioxide. Orange-brown fumes are evolved:

$$KBr(s) + H_2SO_4(l) \rightarrow KHSO_4(s) + HBr(g)$$

$$2HBr(g) + H_2SO_4(l) \rightarrow Br_2(g) + SO_2(g) + 2H_2O(l)$$

Exam tip

The oxidation number of fluorine in all its compounds is −1.

Exam tip

The presence of the acidic HCl can be tested by dipping a glass rod in concentrated ammonia and placing it where the acidic fumes leave the test tube. White smoke (of ammonium chloride) is seen.

Iodide is an even stronger reducing agent and can reduce sulfuric acid further. The reaction produces purple fumes of iodine, a smell of rotten eggs from the hydrogen sulfide and a brown sludge in the test-tube.

The different reduction products of chlorides, bromides and iodides with concentrated sulfuric acid clearly show the increasing power of the halide ions as reducing agents.

- Cl^- does not reduce H_2SO_4.
- Br^- reduces it from the +6 to the +4 state (SO_2).
- I^- reduces it to the −2 state, H_2S.

Reaction with silver nitrate

The test solution is made acidic with nitric acid, which decomposes carbonates or sulfites that would interfere with the test. Silver nitrate solution is then added:

$$Ag^+(aq) + Cl^-(aq) \rightarrow AgCl(s) \quad \text{white precipitate}$$

$$Ag^+(aq) + Br^-(aq) \rightarrow AgBr(s) \quad \text{cream precipitate}$$

$$Ag^+(aq) + I^-(aq) \rightarrow AgI(s) \quad \text{yellow precipitate}$$

The precipitates are then treated with ammonia solution. Silver chloride dissolves in dilute ammonia to give a colourless solution:

$$AgCl(s) + 2NH_3(aq) \rightarrow [Ag(NH_3)_2]^+(aq) + Cl^-(aq)$$

Silver bromide dissolves in concentrated but not in dilute ammonia to give a colourless solution:

$$AgBr(s) + 2NH_3(aq) \rightarrow [Ag(NH_3)_2]^+(aq) + Br^-(aq)$$

Silver iodide is too insoluble to react with ammonia.

Silver halides are decomposed by light. Photochromic lenses contain silver chloride, which dissociates to form silver in the glass. When the light source is removed, the silver recombines with the chlorine.

Exam tip

This reaction is important for observing the hydrolysis of halogenoalkanes, such as $CH_3CHBrCH_3$ (see the student guide covering Topics 6–10 in this series).

Reactions of hydrogen halides

With water

In aqueous solution the hydrogen halides form strongly acidic solutions. The formation of the hydrated H_3O^+ and X^- ions (X = Cl, Br or I) liberates enough energy to compensate for the breaking of the H–X bond. HI is the strongest acid of the hydrogen halides as the H–I bond is the weakest and so the easiest to break.

$$HX(g) + H_2O(l) \rightarrow H_3O^+(aq) + X^-(aq)$$

With ammonia

The solutions are typical strong acids and react with bases to form salts. Thus with ammonia:

$$HX(aq) + NH_3(aq) \rightarrow NH_4X(aq)$$

Fluorine and astatine

The chemistry of the halogens shows a moderately clear trend from fluorine to iodine, so knowledge of the chemistry of chlorine, bromine and iodine enables the chemistry of the other halogens, fluorine and astatine, to be predicted.

Fluorine

It would be expected that, based on the chemistry of the other halogens, fluorine has the following properties:

■ It is a powerful oxidising agent — the most powerful, in fact. Fluoride ions cannot be oxidised by any other oxidising agent, so fluorine can only be made by electrolysis.
■ Fluorine is the most electronegative element.
■ It is also the most reactive element and forms compounds with all other elements, apart from the first three noble gases (He, Ne and Ar).
■ Fluorine reacts with metals to give ionic compounds; it reacts with non-metals to give covalent compounds.
■ Fluorine usually brings out the highest possible oxidation state in the element with which it combines — for example, in reaction with sulfur it forms SF_6, whereas reaction of sulfur with chlorine gives SCl_4.

The high reactivity of fluorine is a result of:

■ the weakness of the F–F bond; it is so short that non-bonding electrons in the two atoms repel one another and weaken the bond, which is not much stronger than that in iodine
■ strong ionic bonds formed with metals because the F^- ion is the smallest anion
■ strong covalent bonds formed with non-metallic elements because the bond is short

Astatine

Astatine is radioactive and occurs in vanishingly small traces in the ore uranite. Its longest-lived isotope is astatine-211 (^{211}At), which has a half-life ($t_{1/2}$) of 8.3 hours. Microgram quantities have been synthesised, but the element has never been seen. Tracer experiments using carrier compounds of iodine suggest that astatine is like iodine, but more metallic and would appear as a dark-coloured solid with a metallic sheen. Astatide ions would be easily oxidised to the element and its compounds would show a high degree of covalent character.

Summary

After studying this topic on group 7 you should be able to:

■ describe the colour of the halogens and their solutions
■ write half and overall equations for the oxidising reactions of the halogens with metals, non-metals, Fe^{2+} ions and other halide ions
■ define disproportionation and write equations for the disproportionation of chlorine and of ClO^- ions
■ state the observations when concentrated sulfuric acid is added to solid halides and write equations for chlorides and bromides
■ write equations and describe the colour of the precipitate for the reaction of silver ions with halide ions

Topic 4C Analysis of inorganic compounds

Cations

Groups 1 and 2

These are identified by a flame test (see page 35).

Ammonium ion

Heat the solid or solution with aqueous sodium hydroxide and test the gas evolved:

$$NH_4^+ + OH^- \rightarrow NH_3 + H_2O$$

- *Damp* red litmus turns blue in the ammonia gas given off.
- When a glass rod is dipped in concentrated hydrochloric acid and held in the gas, white smoke of ammonium chloride is observed:

$$NH_3(g) + HCl(g) \rightarrow NH_4Cl(s)$$

Anions

Halides

Add dilute nitric acid followed by aqueous silver nitrate to a solution of the suspected halide (see page 39).

- Chlorides give a white precipitate (of silver chloride), which is soluble in dilute ammonia:

$$Cl^-(aq) + Ag^+(aq) \rightarrow AgCl(s)$$

then

$$AgCl(s) + 2NH_3(aq) \rightarrow [Ag(NH_3)_2]^+(aq) + Cl^-(aq)$$

- Bromides give a cream precipitate (of silver bromide), which is insoluble in dilute ammonia but soluble in concentrated ammonia.
- Iodides give a pale yellow precipitate (of silver iodide), which is insoluble in both dilute and concentrated ammonia.

Carbonates and hydrogencarbonates

Add dilute hydrochloric acid to the solid (or to the solution). Both give off a gas that turns limewater milky:

$$2H^+ + CO_3^{2-} \rightarrow H_2O + CO_2$$

$$H^+ + HCO_3^- \rightarrow H_2O + CO_2$$

Sulfates

To a solution of the suspected sulfate add dilute hydrochloric acid followed by barium chloride solution. Sulfates give a white precipitate of barium sulfate:

$$SO_4^{2-}(aq) + Ba^{2+}(aq) \rightarrow BaSO_4(s)$$

■ Topic 5 Formulae, equations and amounts of substance

The abilities to calculate formulae from data, write correct formulae, balance equations and perform quantitative calculations are fundamental to the whole study of chemistry. This topic introduces these ideas, and its content could be examined in any of the papers.

The ability to lay out calculations in a comprehensible manner depends on:
- knowing what you are doing and not being reliant on rote learning of formulae
- realising that calculations need linking words and phrases to make them readable
- the correct use of units *throughout the calculation*

Using units throughout a calculation is advantageous. It means that you are less likely to get calculations wrong because you are more aware of what you are doing. Some of the advantages are:
- an awareness that equations are not merely symbols, but express relationships between physical quantities
- a check on whether the equations used are in fact correct, because of the in-built check on the units of the answer
- a gradual awareness of what sort of magnitude of answer is reasonable in a given set of circumstances

Amount of substance and the Avogadro constant

In chemistry, the term *amount* has a technical meaning. It refers to the number of moles of substance being considered.

A **mole** is that amount of substance that contains the same number of entities (molecules, ions, atoms) as there are atoms in exactly $0.012\,kg$ ($12\,g$) of the isotope ^{12}C. The entities concerned must also be specified.

The number of entities in a mole of substance is called the **Avogadro constant**, N_A. Its value is $6.02214179 \times 10^{23}\,mol^{-1}$.

The mole is used because chemistry takes place between particles (molecules, ions or atoms). However, the most convenient way of measuring materials in the laboratory is to weigh them. The Avogadro constant, together with the idea of relative molecular mass, provides the necessary link. Water, H_2O, has a relative molecular mass of 18.0. Since $12\,g$ of ^{12}C contains (by definition) N_A carbon atoms, it follows that $18\,g$ of water contains N_A water molecules, since every molecule has a mass $18.0/12$ times that of a ^{12}C atom. In fact, for any molecular substance, its relative molecular mass, in grams, contains N_A molecules of that substance. For an ionic substance, its formula mass, in grams, has N_A formula units. This enables reacting masses to be calculated (see page 46).

Exam tip

The number of entities = number of moles × 6.02×10^{23}.

Knowledge check 29

Calculate the number of methane molecules in $0.125\,mol$ of methane.

Exam tip

For an ionic solid, such as NaCl, there are 6.02×10^{23} Na+ ions and 6.02×10^{23} Cl- ions per mole. For $CaCl_2$ there are 6.02×10^{23} Ca^{2+} ions and $2 \times 6.02 \times 10^{23}$ Cl- ions per mole.

Knowledge check 30

Calculate the *total* number of ions in $0.125\,mol$ of aluminium sulfate, $Al_2(SO_4)_3$.

Exam tip

One mole of a substance equals its relative molecular mass in grams.

Molar mass is the mass (in grams) of a substance that contains N_A molecules or, if ionic, N_A formula units, of the substance. Its units are $g\,mol^{-1}$ and it is numerically the same as the relative molecular (formula) mass of the substance.

Empirical and molecular formulae

The empirical formula of a compound is its formula in its lowest terms — the simplest whole number ratio of atoms in the formula. Therefore, CH_2 is the empirical formula of any compound C_nH_{2n}, i.e. alkenes, cycloalkanes and poly(alkenes).

The calculation steps give you:

- the moles of each atom
- the ratio of moles of each atom

Example 1

A compound contains 73.47% carbon, 10.20% hydrogen and the remainder is oxygen. The relative molecular mass is 98. Find the empirical and molecular formulae of the compound.

Answer

Atom	Divide by A_r (= moles)	Divide by smallest (= ratio of moles)	Ratio of atoms
Carbon	$\dfrac{73.47}{12.0} = 6.1225$	$\dfrac{6.1225}{1.020} = 6$	6
Hydrogen	$\dfrac{10.20}{1.0} = 10.20$	$\dfrac{10.20}{1.020} = 10$	10
Oxygen	$\dfrac{16.33}{16.0} = 1.020$	$\dfrac{1.020}{1.020} = 1$	1

The compound has the empirical formula $C_6H_{10}O$. The mass of this is:

$$(6 \times 12) + (10 \times 1) + 16 = 98$$

So the empirical formula is also the molecular formula.

Exam tip

If the compound contains oxygen, the mass of oxygen has to be calculated by subtracting the masses of hydrogen and carbon from the mass of the compound.

Example 2

2.4 g of a compound containing carbon, hydrogen and oxygen only was burnt and produced 5.28 g of carbon dioxide and 2.88 g of water. Calculate the moles of carbon dioxide and water and hence find the empirical formula of the compound.

➔

Knowledge check 31

What is the mass of 1 mol of (a) methane, CH_4 and (b) sodium chloride, $NaCl$?

Exam tip

You must know the relationships:

$$moles = \frac{mass}{molar\ mass}$$

$$mass = moles \times molar\ mass$$

Knowledge check 32

Calculate the number of moles in (a) 2.0 g of methane, CH_4 and (b) 2.0 g of sodium chloride, $NaCl$.

Knowledge check 33

What is the mass of 0.12 mol of sodium hydroxide?

Exam tip

When you use the periodic table, you must be careful to use atomic masses, not atomic numbers. For example, carbon is 12.0 not 6.

Content Guidance

Answer

moles of carbon dioxide = 5.28/44 = 0.12 = moles of carbon

mass of carbon = 0.12 × 12 = 1.44 g

moles of water = 2.88/18 = 0.16

moles of hydrogen = 2 × moles of water = 2 × 0.16 = 0.32

mass of hydrogen = 0.32 × 1 = 0.32 g

mass of oxygen in the compound = mass of compound − (masses of carbon and hydrogen)

= 2.4 − 1.44 − 0.32 = 0.64 g

Atom	Moles	Divide by smallest (= ratio of moles)	Ratio of atoms
Carbon	0.12	$\frac{0.12}{0.04} = 3$	3
Hydrogen	0.32	$\frac{0.32}{0.04} = 8$	8
Oxygen	$\frac{0.64}{16} = 0.04$	$\frac{0.04}{0.04} = 1$	1

The empirical formula is C_3H_8O.

Exam tip

As there are two H atoms per molecule in water, the moles of H atoms = 2 × moles of water.

Example 3

This example shows why you must not do any rounding during the calculation.

A compound contains 39.13% carbon, 52.17% oxygen and 8.70% hydrogen. Calculate the empirical formula.

Answer

Atom	Divide by A_r (= moles)	Divide by smallest (= ratio of moles)	Ratio of atoms
Carbon	$\frac{39.13}{12.0} = 3.26$	$\frac{3.26}{3.26} = 1$	1
Oxygen	$\frac{52.17}{16.0} = 3.26$	$\frac{3.26}{3.26} = 1$	1
Hydrogen	$\frac{8.70}{1.0} = 8.70$	$\frac{8.70}{3.26} = 2.66$	2.66

The ratios are not whole numbers. This means you have to multiply by a small integer, usually 2 or 3, to get the answer. In this case, multiplying by three gives 3, 3, 8, so the empirical formula is $C_3H_8O_3$.

To find the **molecular formula**, which is the formula of one molecule of the substance, you need additional information such as the molar mass of the compound.

Chemical equations

The ability to write equations for chemical reactions is important and needs constant practice. Chemical equations, i.e. symbol equations (not word equations which are not equations at all), are central to the language of chemistry.

When writing an equation check that:

- all the formulae are correct
- the equation balances

Whenever you come across a reaction, you should follow these rules:

- Learn the equation for it.
- Check that it balances and make sure that you understand the reason for the presence and amount of each species.
- Try to visualise what is happening in the reaction vessel.

The visualisation of a reaction can be helped by the use of state symbols. The ones commonly used are:

- (s) for solid
- (l) for liquid
- (g) for gas
- (aq) for aqueous solution

It is not always necessary to use state symbols, and can be impossible. For example the reaction between sodium iodide and concentrated sulfuric acid gives a brown mess that contains several substances in a paste with water and acid. Attempts to use state symbols in such cases are not illuminating. However, in the case of thermochemical equations and in equilibria and redox equations, their use is essential.

Chemical equations are not algebraic constructions — they represent real processes. Do not tack atoms or molecules on here and there to make an equation 'balance'; if it does not describe what really happens, it is wrong.

Ionic equations

Ionic equations have the advantage that they ignore all the species that are not involved in the reaction (i.e. those that are just 'looking on' as the reaction proceeds — hence the name **spectator ions**).

The rules for writing ionic equations are:

1 Write out the equation for the reaction.
2 Separate the *soluble* ionic compounds into their ions.
3 Leave covalent compounds and insoluble ionic compounds as they are.
4 Delete the common (spectator) ions from both sides of the equation.

Knowledge check 34

Write the equation for the reaction of aluminium with hydrochloric acid to form aluminium chloride and hydrogen.

Exam tip

When you write state symbols, make them clear. It is common to find exam scripts where the difference between (s) and (g) is not always decipherable.

Content Guidance

This is illustrated by the reaction between solutions of copper sulfate and sodium hydroxide:

$$CuSO_4(aq) + 2NaOH(aq) \rightarrow Cu(OH)_2(s) + Na_2SO_4(aq)$$

$$Cu^{2+}(aq) + \mathbf{SO_4^{2-}(aq)} + \mathbf{2Na^+(aq)} + 2OH^-(aq) \rightarrow Cu(OH)_2(s) + \mathbf{SO_4^{2-}(aq)} + \mathbf{2Na^+(aq)}$$

The spectator ions are in bold. Deleting these gives the ionic equation:

$$Cu^{2+}(aq) + 2OH^-(aq) \rightarrow Cu(OH)_2(s)$$

Knowledge check 35

Balance the ionic equation $Sn^{2+} + Fe^{3+} \rightarrow Sn^{4+} + Fe^{2+}$.

Exam tip

Make sure that ionic equations balance for charge.

Reacting masses

Equations can be used to find how much material could be produced from given starting quantities.

Example

Sodium carbonate weighing 5.3 g is reacted with excess hydrochloric acid. Calculate the mass of sodium chloride produced.

Answer

The molar masses of sodium carbonate and of sodium chloride are needed, since the masses of both are involved in the calculation.

$$\text{molar mass of } Na_2CO_3 = [(2 \times 23.0) + 12.0 + (3 \times 16.0)]\,g\,mol^{-1} = 106.0\,g\,mol^{-1}$$

$$\text{molar mass of } NaCl = 35.5 + 23.0\,g\,mol^{-1} = 58.5\,g\,mol^{-1}$$

Then the steps are as follows:

Step 1 Calculate the amount of starting material (sodium carbonate), where amount means moles.

Step 2 Calculate the amount of sodium chloride from the chemical equation.

Step 3 Calculate the mass of sodium chloride.

The equation is:

$$Na_2CO_3 + 2HCl \rightarrow 2NaCl + CO_2 + H_2O$$

$$\text{amount of } Na_2CO_3 = \frac{5.3\,g}{106.0\,g\,mol^{-1}} = 0.050\,mol$$

$$\text{amount of } NaCl = 2 \times 0.050\,mol = 0.10\,mol$$

$$\text{mass of } NaCl = 0.10\,mol \times 58.5\,g\,mol^{-1} = 5.85\,g$$

Exam tip

Stage 2 uses the stoichiometry of the equation. Here 2 mol of sodium chloride are formed from 1 mol of sodium carbonate.

Equations from reacting masses

Knowing the mass of the materials that react also gives information about the equation for the reaction. This process was a preoccupation of chemists for much of the eighteenth and nineteenth centuries, since it is the basis on which formulae are determined.

Example

In a suitable organic solvent, tin metal reacts with iodine to give an orange solid and no other product. In such a preparation, 5.9 g of tin reacted completely with 25.4 g of iodine. What is the formula of the compound produced?

Answer

molar mass of tin = $118.7 \, \text{g mol}^{-1}$

molar mass of iodine = $126.9 \, \text{g mol}^{-1}$

amount of tin = $\dfrac{5.9 \, \text{g}}{118.7 \, \text{g mol}^{-1}} = 0.050 \, \text{mol}$

amount of iodine atoms = $\dfrac{25.4 \, \text{g}}{126.9 \, \text{g mol}^{-1}} = 0.20 \, \text{mol}$

ratio of iodine:tin = 4:1

So, the compound has the empirical formula SnI_4.

The data do not tell us whether the compound is SnI_4, Sn_2I_8 or some other multiple — a molecular mass determination would be necessary to find out. It is actually SnI_4.

Calculations involving gas volumes

The gas law

The ideal gas equation is:

$$pV = nRT$$

where p = the pressure in Pa, V = the volume in m^3, n = the number of moles, R is the gas constant = $8.31 \, \text{J K}^{-1} \, \text{mol}^{-1}$ and T is the kelvin temperature $(x°C = (273 + x) \, \text{K})$.

This equation allows the volume of a particular gas of known mass to be calculated at any temperature and pressure. It also can be used to find the molar mass of volatile liquids.

Finding the molar mass of a volatile liquid

A sample of the unknown is weighed accurately and then vaporised at a known temperature. Its volume and the pressure are measured. The values are substituted into the ideal gas equation and the number of moles calculated. The molar mass can then be calculated using moles = mass/molar mass.

Exam tip

The units must be as above. The necessary conversions are:

$1 \, cm^3 = 1 \times 10^{-6} \, m^3$

$1 \, dm^3 = 1 \times 10^{-3} \, m^3$

$100 \, \text{kPa} = 100\,000 \, \text{Pa}$

> **Example**
>
> 0.284 g of the same liquid hydrocarbon was injected into a gas syringe. This was then placed in an oven at 150°C where it fully vaporised. Its volume was 84.3 cm³ and the pressure was 106 kPa. Calculate the molar mass of the hydrocarbon.
>
> **Answer**
>
> $$p = 106\,\text{kPa} = 106\,000\,\text{Pa}$$
>
> $$V = 84.3\,\text{cm}^3 = 8.43 \times 10^{-5}\,\text{m}^3$$
>
> $$T = 150 + 273 = 423\,\text{K}$$
>
> $$R = 8.31\,\text{J}\,\text{K}^{-1}\,\text{mol}^{-1}$$
>
> $$n = \frac{pV}{RT} = \frac{106\,000 \times 8.43 \times 10^{-5}}{8.31 \times 423} = 0.00254\,\text{mol}$$
>
> $$\text{molar mass} = \frac{\text{mass}}{\text{moles}} = \frac{0.284\,\text{g}}{0.00254\,\text{mol}} = 112\,\text{g}\,\text{mol}^{-1}$$

Molar volume of a gas

The molar volume of any gas at a given temperature and pressure is the same. It does not depend on the nature of the gas. At s.t.p., i.e. 0°C and 100 kPa pressure, the volume is 22.71 dm³ mol⁻¹. There is no loss of principle in taking it to be 24 dm³ mol⁻¹ at 'room temperature and pressure', but don't forget that this is an approximation.

> ### Core practical 1: Measuring the molar volume of a gas
>
> The molar volume of a gas such as carbon dioxide can be found as follows:
> 1 Weigh a few pieces of marble chips on a top-pan balance.
> 2 Put 50 cm³ of dilute hydrochloric acid in a stoppered flask fitted with a side arm and weigh it.
> 3 Connect the side arm to a delivery tube so that all the gas evolved will be collected over water in a 250 cm³ measuring cylinder.
> 4 Tip the marble chips into the acid and quickly stopper the flask.
> 5 When the measuring cylinder is almost full of gas, disconnect the flask from the delivery tube and immediately weigh it.
> 6 Record the temperature of the water and the atmospheric pressure in the laboratory.

Example

mass of marble chips = 6.45 g

mass of beaker plus acid = 120.13 g

mass of beaker and all contents after reaction = 126.16 g

volume of gas made at room temperature and pressure = 230 cm^3

room temperature = 20°C; room pressure = 1 atm

Answer

mass of carbon dioxide produced = (120.13 + 6.45) − 126.16 = 0.42 g

$$\text{moles of carbon dioxide} = \frac{\text{mass}}{\text{molar mass}} = \frac{0.42\,g}{44\,g\,mol^{-1}} = 0.00955\,mol$$

$$\text{molar volume at r.t.p} = \frac{230\,cm^3}{0.00955\,mol} = 24\,000\,cm^3\,mol^{-1} = 24\,dm^3\,mol^{-1}$$

Calculations involving molar volumes

The molar volume of a gas can be used to find the volumes of gases obtained from any particular reaction.

Exam tip

You must know:

$$\text{moles of gas} = \frac{\text{volume}}{\text{molar volume}}$$

So, at room temperature and pressure:

$$\text{moles of gas} = \frac{\text{volume in dm}^3}{24}$$

Or:

$$\text{moles of gas} = \frac{\text{volume in cm}^3}{24\,000}$$

Knowledge check 36

Calculate the volume of gas, at room temperature and pressure, that contains 0.0246 mol.

Example 1

This is an extension of the reaction considered on page 46.

Sodium carbonate weighing 5.3 g was reacted with excess hydrochloric acid. Calculate the mass of sodium chloride produced and the volume of carbon dioxide obtained at room temperature and pressure.

→

Content Guidance

Answer

The first steps are as before:

$$Na_2CO_3 + 2HCl \rightarrow 2NaCl + CO_2 + H_2O$$

$$\text{amount of } Na_2CO_3 = \frac{5.3\,g}{106.0\,g\,mol^{-1}} = 0.05\,mol$$

$$\text{amount of NaCl} = 2 \times 0.05\,mol = 0.10\,mol$$

$$\text{mass of NaCl} = 0.10\,mol \times 58.5\,g\,mol^{-1} = 5.85\,g$$

The volume of CO_2 produced is given by the amount multiplied by the molar volume:

$$\text{amount of } CO_2 = 0.05\,mol$$

$$\text{volume of } CO_2 = 0.05\,mol \times 24\,dm^3\,mol^{-1} = 1.2\,dm^3$$

> **Exam tip**
>
> The equation shows that 2 mol of NaCl are produced from 1 mol of Na_2CO_3, so the moles of NaCl will be twice the moles of Na_2CO_3.

Example 2

$50\,cm^3$ of propane is burnt in excess oxygen. What volume of oxygen reacts, and what is the volume of carbon dioxide produced?

Answer

In this case, the substances whose volumes are required are gaseous and the initial volume of propane is given. The volumes are in the same ratio as the amounts of each substance in the equation:

$$C_3H_8(g) + 5O_2(g) \rightarrow 3CO_2(g) + 4H_2O(l)$$

$$\text{molar ratio of the gases propane:carbon dioxide} = 1:3$$

$$\text{volume of propane} = 50\,cm^3$$

So:

$$\text{volume of carbon dioxide produced} = (3/1) \times 50 = 150\,cm^3$$

It is not necessary to use the molar volume of the gas.

> **Exam tip**
>
> This method can only be used if both substances in the question are gases.

Example 3

$27\,g$ of propane is burnt in excess oxygen. What volume of oxygen reacts and what is the volume of carbon dioxide produced?

The molar volume of any gas at the temperature and pressure of the experiment is $24\,dm^3\,mol^{-1}$.

The molar mass of propane is $44.0\,g\,mol^{-1}$.

Answer

$$C_3H_8(g) + 5O_2(g) \rightarrow 3CO_2(g) + 4H_2O(l)$$

amount of propane $= \dfrac{27.0\,g}{44.0\,g\,mol^{-1}} = 0.614\,mol$

amount of oxygen $= 0.614\,mol \times 5 = 3.07\,mol$

volume of oxygen $= 3.07\,mol \times 24\,dm^3\,mol^{-1} = 73.7\,dm^3$

The volume of carbon dioxide produced is calculated thus:

amount of carbon dioxide $= 3 \times 0.614\,mol = 1.842\,mol$

volume of carbon dioxide $=$ amount $\times$ molar volume

$= 1.842\,mol \times 24\,dm^3\,mol^{-1}$

$= 44.2\,dm^3$

Exam tip

If the volume of *air* had been asked, the $73.7\,dm^3$ would have to be multiplied by 5, as air is only one-fifth oxygen.

Titration calculations

Concentration

The concentration of a solution can be expressed in several ways, each having its own purpose.

The commonest method is to use **mol dm^{-3}**, where the concentration is given as the amount of solute per dm^3 of *solution*. When solutions are made up, the required amount of solute is measured out and the solvent is then added until the required volume of solution is obtained. It is not possible to predict the volume change when two materials are mixed to give a solution because it depends on how the particles fit together. For example, mixing 50 cm^3 of water and 50 cm^3 of ethanol gives only 97 cm^3 of aqueous ethanol.

It is possible to calculate the mass of a substance in a known volume of a solution of known concentration, as shown below for 50 cm^3 of sodium hydroxide solution of concentration 0.130 mol dm^{-3}. Sodium hydroxide has a molar mass of 40 g mol^{-1}.

amount of NaOH $= 0.050\,dm^3 \times 0.130\,mol\,dm^{-3} = 6.5 \times 10^{-3}\,mol$

mass of NaOH $= 6.5 \times 10^{-3}\,mol \times 40\,g\,mol^{-1} = 0.26\,g$

Exam tip

You must know:

moles = concentration × volume in dm^3

volume in dm^3 = $\dfrac{\text{volume in cm}^3}{1000}$

Knowledge check 37

Calculate the mass of sodium hydroxide required to make 250 cm^3 of a 0.100 mol dm^{-3} solution.

In solutions that are very dilute and in gases such as air containing some airborne pollutants, the amount of material present is small. In such cases the concentration of the solute is given in **parts of solute per million parts of solvent**, by mass. This is abbreviated to **ppm**. A solution having 1 ppm of solute would have 10^{-6} g of solute per gram of solution.

Acid–base titrations: volumetric analysis

Techniques used in volumetric analysis

Volumetric analysis, or titration, involves the reaction of two solutions. The volume and the concentration are known for one solution. For the other solution, only the volume is known. The apparatus used is a burette, a pipette and a graduated (volumetric) flask. Accurate results require careful technique in the use of each of these.

Using a pipette

The use of a pipette filler is obligatory. In safety questions in examinations, credit is not given for answers referring to the use of lab coats, safety glasses or pipette fillers, since all are assumed to be part of routine good practice.

- Using a pipette filler, draw a little of the solution to be pipetted into the pipette and use this to rinse the pipette; discard these rinsings.
- Fill the pipette to about 2–3 cm above the mark. Pipette fillers are difficult to adjust accurately, so quickly remove the filler and close the pipette with your forefinger (not thumb). Release the solution until the bottom of the meniscus is on the mark.
- Immediately transfer the pipette to the conical flask in which you will do the titration and allow the solution to dispense under gravity. Under no circumstances blow it out. Good analysis requires patience. When all the solution appears to have been dispensed, wait 15–20 seconds, touch the tip of the pipette under the surface of the liquid and then withdraw the pipette. It is calibrated to allow for the liquid remaining in the tip.

Using a burette

A burette dispenses solutions accurately only if used correctly — there are two particular pitfalls that can cause serious inaccuracies.

- Making sure that the tap is shut, add about 10 cm³ of the solution you intend to use into the burette and rinse it out with this solution, not forgetting to open the tap and rinse the jet.
- Close the tap and fill the burette using a small funnel. Remove the funnel; titrating with a funnel in the burette can lead to serious error if a drop of liquid in the funnel stem falls into the burette during the titration.
- Bring the meniscus onto the scale by opening the tap over a suitable receptacle.
- Make sure that the burette is full to the tip of the jet. Failure to ensure this is another source of serious error.

Exam tip

The error in using a school pipette is about ± 0.1 cm³. When using a 25 cm³ pipette this can cause a percentage error of 0.1 × 100/25.0 = 0.4%. For a 10 cm³ pipette the error rises to 1%.

Exam tip

Never use a measuring cylinder to measure an accurate volume of a solution.

Exam tip

There is no particular reason to bring the meniscus exactly to the zero mark.

Exam tip

The error when using a burette is $\pm0.05\,cm^3$ per reading. Thus the total error in a titre is $2 \times 0.05 = 0.10\,cm^3$. Making sure that the titre is around $25\,cm^3$ gives an error of around $0.1 \times 100/25 = 0.4\%$. A titre of $12.5\,cm^3$ would have a percentage error of 0.8%.

■ Titration is a two-handed process. Add a suitable indicator to the solution in the conical flask, then swirl this under the burette with your right hand while manipulating the burette tap with your left. Your thumb and forefinger should encircle the burette. This feels awkward at first, but after practice becomes natural and gives good control of the tap.

■ Add the titrant (the solution in the burette) slowly, swirling the flask all the time. As the end point is approached, the indicator will change colour more slowly. The titrant should be added drop by drop near to the end point — your aim is to make it change with the addition of one drop of titrant. Wait a few moments before reading the burette — this is to allow the solution time to drain down the walls of the burette.

■ The titration is repeated until you have two or three concordant titres, that is, volumes that are within $0.2\,cm^3$ or better. Taking the mean of the titres that differ by more than $0.2\,cm^3$ cannot give accurate answers.

Errors

The error in a pipette is $\pm 0.1\,cm^3$. For a $25\,cm^3$ pipette, this is a 0.4% error.

In a burette it is $\pm0.05\,cm^3$ *per reading*, so an error of $\pm0.1\,cm^3$ per titre.

A normal balance has an error of $\pm0.01\,g$, so an error of $\pm0.02\,g$ per weight of solid taken.

To calculate the total error, the percentage error for each measurement must be worked out. These are then added to give the total possible percentage error in the final result.

Indicators

The two common indicators for acid–base titrations are methyl orange and phenolphthalein. Methyl orange is red below pH 3 and yellow above about pH 6. Phenolphthalein is colourless below about pH 8 and pink (or magenta) above pH 10.

Many acid–base titrations produce a large change in pH from about 1.5–2 to 12–13 (if base is being added to acid) within a few drops of the end point volume, so either indicator can be used. However, a strong acid–weak base titration needs methyl orange and a strong base–weak acid titration needs phenolphthalein. (The theory of indicators is covered in the second year of the A level course.) Another example of an indicator is bromothymol blue, which is yellow below pH 6 and blue above pH 8.

Volumetric calculations

It is essential to lay out calculations clearly. They must be easy to read, make sense and be realistic in terms of significant figures used. Adopting the following principles will improve your understanding of volumetric calculations.

Exam tip

percentage error in a measurement = error × 100/the measurement

Knowledge check 38

Calculate the total percentage error in a titration that involves weighing out $1.23\,g$ of a solid and doing a titration using a $25\,cm^3$ pipette and with a mean titre of $21.40\,cm^3$.

Exam tip

Do not say that you add $250\,cm^3$ of water to the solid, as this will not accurately make $250\,cm^3$ of solution.

- The whole calculation uses numbers of moles rather than some formula that might be mis-remembered or misapplied.
- Use units throughout.
- The word 'amount' is used in its technical chemical sense, i.e. a quantity of moles.
- The molar mass of a compound is calculated and written down. Incorrect molar masses cost candidates many exam marks because the examiner cannot tell whether the error is arithmetical or chemical.
- The dimensionless quantity *relative* molecular mass is not used. No calculation uses this quantity.

Example

$25.0\,cm^3$ of a solution of sodium carbonate was titrated with hydrochloric acid solution of concentration $0.108\,mol\,dm^{-3}$ using methyl orange indicator. The volume needed was $27.2\,cm^3$. Find the concentration of the sodium carbonate solution in $mol\,dm^{-3}$ and in $g\,dm^{-3}$ of the anhydrous salt.

The reaction is:

$$Na_2CO_3 + 2HCl \rightarrow 2NaCl + H_2O + CO_2$$

amount of hydrochloric acid used = $0.0272\,dm^3 \times 0.108\,mol\,dm^{-3} = 0.00294\,mol$

Thus:

amount of sodium carbonate = $\frac{1}{2} \times 0.00294\,mol = 0.00147\,mol$ in $25.0\,cm^3$

Thus:

concentration of sodium carbonate solution = $0.00147\,mol/0.025\,dm^3$

$$= 0.0588\,mol\,dm^{-3}$$

The molar mass of anhydrous sodium carbonate is:

$$[(2 \times 23) + 12 + (3 \times 16)]\,g\,mol^{-1} = 106\,g\,mol^{-1}$$

The concentration of the sodium carbonate is therefore:

$$0.0588\,mol\,dm^{-3} \times 106\,g\,mol^{-1} = 6.23\,g\,dm^{-3}$$

Exam tip

Remember:

moles = volume in dm^3 × concentration

Knowledge check 39

$25.0\,cm^3$ of a $0.0500\,mol\,dm^{-3}$ solution of ethanedioic acid was neutralised by $23.4\,cm^3$ of a sodium hydroxide solution. Calculate the concentration, in $mol\,dm^{-3}$, of the sodium hydroxide.

$$H_2C_2O_4 + 2NaOH \rightarrow Na_2C_2O_4 + 2H_2O$$

Percentage yields and atom economies

Until recently, the effectiveness of a synthetic process in chemistry was judged by the percentage yield of the desired compound. Most reactions do not give a quantitative yield — there may be competing reactions, and there will be handling losses unless great care is taken. The percentage yield is given by:

$$\% \text{ yield} = \frac{\text{actual yield of product} \times 100}{\text{calculated yield of product}}$$

With percentage yield, the focus is on the product; the amount of other, possibly useless, materials is not taken into account.

A different measure of the effectiveness of a synthetic process was put forward by Trost (1998). This is the idea of the **atom economy** of a reaction. Atom economy is defined as:

$$\text{atom economy} = \frac{\text{molar mass of the product} \times 100}{\text{sum of the molar masses of all the reactants}}$$

Knowledge check 40

The solvent ethyl ethanoate, $CH_3COOC_2H_5$, can be prepared in two ways:

$$CH_3COCl + C_2H_5OH \rightarrow CH_3COOC_2H_5 + HCl$$

$$CH_3COOH + C_2H_5OH \rightarrow CH_3COOC_2H_5 + H_2O$$

Explain which has the higher atom economy.

The better the atom economy, the greater are the proportions of the starting materials that finish up in the product. The advantages of processes with high atom economies are:

- The consumption (and therefore the cost) of starting materials is minimised.
- There is less production of useless materials in the reaction. This becomes increasingly important as the costs of waste disposal and of compliance with stringent disposal regulations rise rapidly.
- Energy costs are often lower.

Reactions can be categorised according to their atom economies:

- Addition reactions have the highest atom economies since all the reactants are consumed in making the product. Examples include addition polymerisation, addition of HBr to alkenes and the hydrogenation of C=C bonds.
- Substitution reactions have lower atom economies, since one group in a molecule is displaced by another, and the displaced material may be of little use. Examples include the reaction, in sulfuric acid solution, of sodium bromide with an alcohol to form a bromoalkane. The products include dilute sodium sulfate, which has no commercial value and is discarded. It is not toxic or dangerous — just useless.
- Elimination reactions have the lowest atom economies since they produce at least two materials from one, and one of these is likely to be of little use. An example is the elimination, using KOH in an ethanol solvent, of HBr from a bromoalkane to give an alkene. The products include potassium bromide in dilute aqueous solution — not particularly problematic, but it does have to be disposed of.

Summary

After studying this topic, you should be able to:

■ define relative atomic mass, relative isotopic mass and atom economy
■ calculate the moles of a substance given its mass
■ calculate the moles of a gaseous substance given its volume and the molar volume
■ calculate the moles of a solute given its concentration and the volume of the solution
■ understand the term parts per million (ppm)
■ calculate the empirical formulae from % composition or combustion data

■ calculate the mass (or volume for gases) of product given the equation for the reaction and the mass (or volume) of a reactant
■ describe how to prepare a standard solution
■ calculate the concentration of a solution given titration data
■ calculate the % errors involved in weighing and using pipettes and burettes
■ calculate the % yield given the mass of a reactant and the product
■ calculate the atom economy given the equation

■ Practical aspects

Questions will be asked about practical chemistry in both AS papers 1 and 2, and especially in paper 3 of the A level examination.

Standard solutions

The solution for which the concentration is accurately known is the **standard solution**. The concentration may have been found by a previous titration or by weighing the solute and making it to a known volume. A standard solution made by weighing is a **primary standard solution**.

Core practical 2: Preparing a primary standard solution

In this experiment a standard primary solution has to be prepared. The substance used must:
■ be able to be obtained pure
■ not lose water of crystallisation to the air
■ not absorb water or carbon dioxide from the air
A suitable substance is ethanedioic acid, $H_2C_2O_4.2H_2O$.

Exam tip

The error when using a top-pan balance is ±0.01 g per reading. Thus the total error may be $2 \times 0.01\,g = 0.02$. For a mass of 2.34 g the percentage error may be $0.02 \times 100/2.34 = 0.85\%$.

Method
■ Make sure the balance pan is clean and dry. Place the weighing bottle on the pan and tare the balance.
■ Add a suitable amount of the solid to the bottle. Do this by taking the bottle off the pan and adding the solid away from the balance, so any spillage does not fall on the pan. Try to avoid spillage.

- When you have the required amount weigh it and write its value down immediately in your notebook. Tip the contents into a beaker and weigh the bottle again:

  mass of solid = weight of (bottle + solid) – weight after emptying solid

  This makes sure that you have not counted any solid left in the weighing bottle.
- Having transferred the solid to a beaker, add about $50\,cm^3$ of water. The solution is stirred with a glass rod until the solid has dissolved. It is transferred completely to the graduated flask by pouring it down the same glass rod into a funnel and by washing any remaining solution in the beaker and on the glass rod into the funnel.
- Add pure water so that the lower level of the meniscus is on the mark. Stopper the flask and mix thoroughly by inverting and shaking the flask vigorously five or six times. Simple shaking is not enough; most serious errors in volumetric analysis can be traced to poor mixing of the solution in the graduated flask.

Knowledge check 41

Calculate the mass of hydrated ethanedioic acid required to make $250\,cm^3$ of a $0.0500\,mol\,dm^{-3}$ solution.

Exam tip

Do not say that you add $250\,cm^3$ of water to the solid, as this will not accurately make $250\,cm^3$ of solution.

Core practical 3: Finding the concentration of a solution of hydrochloric acid

A secondary standard solution, such as sodium hydroxide, is made up as accurately as possible using the same method as above. It is then standardised by titration against the primary standard solution of ethanedioic acid, and its concentration calculated.

Calculation of the concentration of the secondary standard solution

The route is as follows:

volume of primary standard solution → moles of primary standard solution → moles of secondary standard solution → concentration of secondary standard solution

Note that the equation for ethanedioic acid reacting with sodium hydroxide is:

$$H_2C_2O_4 + 2NaOH \rightarrow Na_2C_2O_4 + 2H_2O$$

So:

moles of NaOH = 2 × moles of $H_2C_2O_4$

The sodium hydroxide used in core practical 3 is now itself a standard solution and can be used to find the concentration of any acid.

The titration method and the calculation are exactly the same as on page 54.

Other practical requirements

For AS paper 1 and for A level paper 3, the following need to be known:

- Analysis of cations and anions — see page 41.
- Measurement of the molar mass of a volatile liquid — see page 47.
- Measurement of the molar volume of a gas (core practical 1) — see page 48.
- Titration techniques and errors — see pages 52–53.

Hazard and risk

Hazard is the **potential** to do harm.

Risk is the **probability** of harm occurring.

Possible hazards of a substance include:

- toxicity (e.g. lead compounds)
- absorption through the skin (e.g. 1-bromobutane)
- irritation if inhaled (e.g. chlorine)
- corrosive compounds (e.g. concentrated sulfuric acid)
- high flammability (e.g. ethoxyethane, $C_2H_5OC_2H_5$)
- carcinogenic compounds (e.g. benzene, C_6H_6)

A reaction might be hazardous if it is rapid and highly exothermic or if it produces a volatile hazardous product.

The risk relates to how a hazardous material is used. For instance, using small quantities or taking specific precautions (e.g. wearing gloves or using a water bath for heating) makes the probability of harm occurring less and so reduces the risk. Hazardous materials used in small quantities with proper containment pose little risk.

> **Exam tip**
>
> When answering a question about hazard and risk, *never* give standard safety precautions, such as wearing safety glasses and a laboratory coat, as methods of minimising risk.

Summary

You must know how to:
- prepare a primary standard solution of ethanedioic acid
- make a secondary standard solution of sodium hydroxide
- find the molar mass of a liquid and a gas

You must be able to describe:
- a flame test and the test for ammonium ions
- tests for carbonate, halide and sulfate ions

Check that you:
- know the difference between hazard and risk
- can identify hazards of specific compounds and how to minimise the risks when using them in the laboratory

Questions & Answers

This section contains multiple-choice and structured questions similar to those you can expect to find in AS paper 1 and parts of questions in the A level examinations. The questions given here are not balanced in terms of types of question or level of demand — they are not intended to typify real papers, only the sorts of questions that could be asked.

The answers given are those that would be expected from a top-grade student. They are not 'model answers' to be regurgitated without understanding. In answers that require more extended writing, it is usually the ideas that count rather than the form of words used; the principle is that correct and relevant chemistry scores.

Comments

Comments on the questions are indicated by the icon ⓔ. They offer tips on what you need to do to get full marks. Responses to questions might be followed by comments, preceded by the icon ⓔ, that explain the correct answer, point out common errors made by students who produce work of C-grade or lower, or contain additional material that could be useful to you.

The exam papers

Each AS paper lasts 1 hour 30 minutes and is worth 80 marks. A minimum of 20 marks will be awarded for the understanding of experimental skills and the interpretation of quantitative and qualitative experimental data. There will be 10–20 multiple-choice questions embedded in structured questions. In the A level exams, questions on some or all of the topics in this book will also be asked in papers 1 and 2 (both 1 hour 45 minutes, with a maximum of 90 marks) and in paper 3 (2 hours 30 minutes, with a maximum of 120 marks).

Assessment objectives

The AS and A level exams have three assessment objectives (AOs). The percentages of each are very similar in the AS and the A level papers, with a slightly heavier weighting of AO3 in the A level.

AO1 is 'knowledge and understanding of scientific ideas, processes and procedures'. This makes up 36% of the exam. You should be able to:

- recognise, recall and show understanding of scientific knowledge
- select, organise and present information clearly and logically, using specialist vocabulary where appropriate

AO2 is 'application of knowledge and understanding of scientific ideas, processes and procedures'. This makes up 42% of the exam. You should be able to:

- analyse and evaluate scientific knowledge and processes
- apply scientific knowledge and processes to unfamiliar situations
- assess the validity, reliability and credibility of scientific information

AO3 is 'analysis, interpretation and evaluation of scientific ideas and data'. This makes up 22% of the exam. You should be able to make scientific judgements, reach conclusions and evaluate and improve described practical procedures.

Command terms

The following command terms are used in the specification and in the AS and A level papers. You must distinguish between them carefully.

- **Give, state or name** — no explanation is needed.
- **Identify** — give the name or formula.
- **Define/state what is meant by** — give a simple definition, without any explanation.
- **Describe** — state the characteristics of a particular method or process. No explanations are needed.
- **Explain** — use chemical theories or principles to say why a particular property of a substance or series of substances is as it is. It requires the making and justification of one or more points.
- **Deduce** — draw conclusions from information provided.
- **Calculate** — you are advised to show your working, so that consequential marks can be awarded even if you made a mistake in an earlier part of the calculation. The answer should include units and should be written to three significant figures unless the question asks for a specific or a suitable number of significant figures.

Revision

Start your revision in plenty of time. Make a list of what you need to do, emphasising the things that you find most difficult, and draw up a detailed revision plan. Work back from the examination date, ideally leaving an entire week free from fresh revision before that date. Be realistic in your revision plan and then add 25% to the timings because everything takes longer than you think.

When revising:

- make a note of difficulties and ask your teacher about them — if you do not make such notes, you will forget to ask
- make use of past papers, but remember that these may have been written to a different specification
- revise ideas, rather than forms of words — you are after *understanding*
- remember that scholarship requires time to be spent on the work

When you use the example questions in this book, make a determined effort to answer them before looking up the answers and comments. Remember that the answers given here are not intended as model answers to be learnt parrot-fashion. They are answers designed to illuminate chemical ideas and understanding.

The exam paper

- Read the question. Questions usually change from one examination to the next. A question that looks the same, at a cursory glance, as one that you have seen before usually has significant differences when read carefully. Needless to say, you will not receive credit for writing answers to your own questions.

- Be aware of the number of marks available for a question. That is an excellent pointer to the number of things you need to say.
- Do not repeat the question in your answer. The danger is that you fill up the space available and think that you have answered the question, when in reality some or maybe all of the real points have been ignored.
- The name of a 'rule' is not an explanation for a chemical phenomenon. For example, in equilibrium, a common answer to a question on the effect of changing pressure on an equilibrium system is 'because of Le Chatelier's principle…'. That is simply a name for a rule — it does not explain anything.

Multiple-choice questions

Answers to multiple-choice questions are machine-marked. Multiple-choice questions need to be read carefully; it is important not to jump to a conclusion about the answer too quickly. You need to be aware that one of the options might be a 'distracter'. An example of this might be in a question having a numerical answer of, say, $-600\,\text{kJ}\,\text{mol}^{-1}$; a likely distracter would be $+600\,\text{kJ}\,\text{mol}^{-1}$.

Some questions require you to think on paper — there is no demand that multiple-choice questions should be carried out in your head. Space is provided on the question paper for rough working. It will not be marked, so do not write anything that matters in this space.

For each of the questions there are four suggested answers, A, B, C and D. You select the best answer by putting a cross in the box beside the letter of your choice. If you change your mind you should put a line through the box and then indicate your alternative choice. Making more than one choice does not earn any marks. Note that this format is not used in the multiple-choice questions in this book.

Each test has at least ten multiple-choice questions, which are embedded in longer questions.

Online marking

It is important that you have some understanding of how examinations are marked, because to some extent it affects how you answer them. Your examination technique partly concerns chemistry and partly must be geared to how the examinations are dealt with physically. You have to pay attention to the layout of what you write. Because all your scripts are scanned and marked online, there are certain things you must do to ensure that all your work is seen and marked. These are covered below.

As the examiner reads your answer, decisions have to be made — is this answer worth the mark or not? Your job is to give the *clearest possible answer* to the question asked, in such a way that your chemical understanding is made obvious to the examiner. In particular, you must not expect the examiner to guess what is in your head; you can be judged only by what you write.

Because examination answers cannot be discussed, you must make your answers as clear as possible. This is one reason why you are expected, for example, to show working in calculations.

It is especially important that you *think before you write*. You will have a space for your answer on the question paper, and that space is what the examiner has judged to be a reasonable amount of space for the answer. Because of differing handwriting sizes, because of false starts and crossings-out, and because some students have a tendency to repeat the question in the answer space, that space is never exactly right for all candidates.

Edexcel exam scripts are marked online, so few examiners will handle a real, original script. The process is as follows:

- When the paper is set it is divided up into items — often, but not necessarily, a single part of a question. These items are also called clips.
- The items are set up so that they display on-screen, with check-boxes for the score and various buttons to allow the score to be submitted or for the item to be processed in some other way.
- After you have written your paper it is scanned; from that point all the handling of your paper is electronic. Your answers are tagged with an identity number.
- It is impossible for an examiner to identify a centre or a candidate from any of the information supplied.

Common pitfalls

The following list of potential pitfalls to avoid is particularly important:

- **Do not write in any colour other than black.** This is now an exam board regulation. The scans are entirely black-and-white, so any colour used simply comes out black — unless you write in red, in which case it does not come out at all. The scanner cannot see red (or pink or orange) writing. So if, for example, you want to highlight different areas under a graph, or distinguish lines on a graph, you must use a different sort of shading rather than a different colour.
- **Do not use small writing.** Because the answer appears on a screen, the definition is slightly degraded. In particular, small numbers used for powers of 10 can be difficult to see. The original script is always available but it takes a relatively long time to get hold of it.
- **Do not write in pencil.** Faint writing does not scan well.
- **Do not write outside the space provided without saying, within that space, where the remainder of the answer can be found.** Examiners only have access to a given item; they cannot see any other part of your script. So if you carry on your answer elsewhere but do not tell the examiner within the clip that it exists, it will not be seen. Although the examiner cannot mark the out-of-clip work, the paper will be referred to the Principal Examiner responsible for the paper.
- **Do not use asterisks or arrows as a means of directing examiners where to look for out-of-clip items.** Tell them in words. Students use asterisks for all sorts of things and examiners cannot be expected to guess what they mean.
- **Do not write across the centre-fold of the paper from the left-hand to the right-hand page.** A strip about 8 mm wide is lost when the papers are guillotined for scanning.
- **Do not repeat the question in your answer.** If you have a question such as 'Define the first ionisation energy of calcium', the answer is 'The energy change per mole for the formation of unipositive ions from isolated calcium atoms in the gas phase'; or, using the equation, 'The energy change per mole for $Ca(g) \rightarrow Ca^+(g) + e^-$'.
Do not start by writing 'The first ionisation energy for calcium is defined as...' because by then you will have taken up most of the space available for the answer. Examiners know what the question is — they can see it on the paper.

■ Structured and multiple-choice questions

In the examinations, the answers are written on the question paper. Here, the questions are not shown in examination paper format.

Question 1

(a) Define the term 'atomic number'. (1 mark)

(b) Sulfur has a radioactive isotope $^{35}_{16}$S. The composition of one 2^- ion of this isotope is:

	Number of protons	Number of neutrons	Number of electrons
A	16	19	16
B	16	19	18
C	19	16	17
D	19	16	19

(1 mark)

e Make sure you know which number refers to the number of protons and which to the number of nucleons.

(c) (i) Write the equation that represents the third successive ionisation energy of an element X. (1 mark)

e Remember to add charges and state symbols.

(ii) The first seven successive ionisation energies of the element X are:

1400   2900   4600   7500   9400   53 000   64 000

The outer electron configuration of element X is:

A $2s^2$

B $2s^2 2p_x^2 2p_y^1$

C $2s^2 2p_x^1 2p_y^1 2p_z^1$

D $2s^2 2p_x^2 2p_y^2 2p_z^1$ (1 mark)

(iii) Explain why the second ionisation energy of X is more endothermic than the first. (2 marks)

e Mention must be made about both the first and the second ionisation energy.

(d) The set of ions in which the members all have the same electron configuration is:

A Fe^{2+}, Fe^{3+}

B N^{3-}, O^{2-}, F^-

C SO_4^{2-}, SeO_4^{2-}, TeO_4^{2-}

D F^-, Cl^-, Br^- (1 mark)

(Total: 7 marks)

Student answer

(a) It is the number of protons in one atom of the element. ✓

(b) B ✓

ⓔ The 2^- ion has two more electrons than protons.

(c) (i) $X^{2+}(g) - e \rightarrow X^{3+}$ ✓

ⓔ You may write – e on the left or + e on the right.

(ii) C ✓

ⓔ There is a big jump between the fifth and sixth values, so there are five outer electrons.

(iii) In the first ionisation energy the electron is being removed from a neutral atom, whereas in the second it is being removed from a positive ion. ✓ So there is greater attraction between the electron leaving and the already 1+ ion ✓ and so more energy is needed.

(d) B ✓

ⓔ These all have the same electronic configuration as neon. The ions in D have the same number of electrons in their outer orbit.

ⓔ **Total score: 7/7 marks**

Question 2

(a) When a sample of copper is analysed using a mass spectrometer, its atoms are ionised, accelerated and then separated according to their mass/charge (m/z) ratio.

   (i) Explain how the atoms of the sample are ionised. (2 marks)

   (ii) State how the resulting ions are accelerated. (1 mark)

   (iii) State how the ions are separated according to their m/z values. (1 mark)

(b) For a particular sample of copper, two peaks were obtained in the mass spectrum, showing an abundance of 69.10% at m/z 63, and 30.90% at m/z 65.

   (i) Give the formula of the species responsible for the peak at m/z 65. (1 mark)

ⓔ A perfect answer would include the atomic number, the mass number and any charge on the species.

   (ii) State why two peaks, at m/z values of 63 and 65, were obtained in the mass spectrum, but no peak is found at an m/z value of 64. (1 mark)

   (iii) Calculate the relative atomic mass of this sample of copper to three significant figures, using the data given above. (2 marks)

(Total: 8 marks)

Student answer

(a) (i) Fast-moving *or* energetic electrons strike the atoms, ✓ removing electrons from the sample atoms, and forming positive ions ✓.

e The copper sample would be on a heated probe; although the vapour pressure of copper is very small, the number of atoms volatilised at the extremely low pressure in the mass spectrometer is large enough to be detected.

(ii) In an electric *or* electrostatic field ✓

(iii) By a magnetic field ✓

(b) (i) $^{65}_{29}Cu^+$ ✓

e Many students forget to put the essential positive charge when giving the formula of an ion detected by a mass spectrometer. $^{65}Cu^+$ would be accepted.

(ii) Because naturally occurring copper contains two isotopes of mass 63 and 65. ✓

(iii) relative atomic mass $= \dfrac{(63 \times 69.10) + (65 \times 30.90)}{69.10 + 30.90} = 63.6$ ✓✓

e The question asks for *three* significant figures. If you give 63.618 or 63.62 you will not score the mark. This is a silly way to lose marks.

e **Total score: 8/8 marks**

Question 3

(a) Explain the following observations in terms of the intermolecular forces present:

(i) Ethanol has a much higher boiling temperature (352 K) than propane (231 K), even though the molecules have the same number of electrons. (4 marks)

e You must mention all the intermolecular forces involved.

(ii) Ethanol is less polar than chloroethane, but ethanol is soluble in water whereas chloroethane is not. (4 marks)

(b) Which of A–D represents most accurately the hydrogen bonding that occurs between ethanol and water? (1 mark)

(Total: 9 marks)

Questions & Answers

Student answer

(a) (i) The molecules have the same number of electrons and so the London forces are similar. ✓ Ethanol also forms intermolecular hydrogen bonds ✓, whereas propane has London forces only ✓, so more energy is required to separate the ethanol molecules ✓.

ⓔ The question is a comparison between two molecules, so both must be referred to in your answer. Weaker answers often ignore one of the molecules.

(ii) Ethanol can form hydrogen bonds with water ✓, which compensates energetically for the breaking of intermolecular hydrogen bonds in water itself ✓. Chloroethane cannot form hydrogen bonds with water since chlorine is too big ✓, so the (endothermic) disruption of intermolecular hydrogen bonds in water cannot be offset by (exothermic) solvent–solute bonding ✓.

ⓔ It is important to recognise that dissolving requires solvent–solvent bonds to be broken and that the energy required to do this has to come from formation of a solvent–solute interaction.

(b) A ✓

ⓔ The O–H–O bond angle must be 180°, so the answer is A not D.

ⓔ Total score: 9/9 marks

Question 4

(a) Explain the following observations. Include details of the *bonding* in, and the *structure* of, each substance.

    (i) The melting temperature of diamond is much higher than that of iodine.   (5 marks)

ⓔ You must state the type of structures of both iodine and diamond, the type of forces between the particles and the relationship between the strength and the melting temperature.

    (ii) Sodium chloride has a high melting temperature (approximately 800°C).   (3 marks)

ⓔ As above, the type of structure and force must be described.

(b) Explain in terms of its structure and bonding why aluminium is a good conductor of electricity.   (3 marks)

(Total: 11 marks)

ⓔ Read the question carefully. You must first answer about the structure and bonding of aluminium and then explain why this allows it to conduct electricity.

Student answer

(a) (i) Iodine is molecular covalent ✓ so has weak intermolecular forces ✓; diamond is giant covalent ✓, so has strong intramolecular covalent bonds throughout the crystal lattice ✓. The stronger the forces, the more energy is needed to overcome them to melt the crystal. ✓

e When discussing the bonding of covalent substances, it is important to distinguish between *inter*molecular forces, which exist between *molecules,* and *intra*molecular forces, which are between *atoms.* Covalent bonds are strong.

(ii) Sodium chloride has a lattice ✓ of oppositely charged ions ✓ with strong attractions throughout the lattice ✓.

e The extension of the attractions throughout the lattice is important. There are no pairs of ions that can be selected over any other pair; the attraction is uniform throughout the crystal. There are also repulsions in the lattice between ions of the same charge; the attraction is a net attraction.

(b) Metal ions are in a lattice ✓ bonded by attraction to delocalised electrons *or* embedded in a sea of electrons ✓. Mobile electrons enable conduction. ✓

e Be sure to state that the electrons are mobile. Just saying that they form a 'sea' is not enough. It is also important to say that the electrons are attracted to the ions.

e **Total score: 11/11 marks**

Question 5

(a) (i) Draw the shape of the H_2O molecule and mark on your drawing the value of the H–O–H angle. (2 marks)

(ii) Explain why water has this shape and bond angle. (3 marks)

e Make sure that you state the number of bonds and lone pairs around the oxygen atom and then explain how this causes the bond angle.

(b) The shape of an ammonia molecule, NH_3, is:

A trigonal planar with three electron pairs around the nitrogen atom

B tetrahedral with four electron pairs around the nitrogen atom

C pyramidal with four electron pairs around the nitrogen atom

D square planar with four electron pairs around the nitrogen atom (1 mark)

(c) Explain the meaning of the term *electronegativity*. (1 mark)

(d) Tetrachloromethane (CCl_4) has polar C–Cl bonds.

(i) Explain why C–Cl bonds are polar. (1 mark)

(ii) Explain whether or not the CCl_4 molecule is polar overall. (2 marks)

(Total: 10 marks)

e Your answer should be about the polarity not the charges.

Student answer

(a) (i)

ⓔ Molecule drawn bent ✓ with a bond angle of 104.5° (but allow between 104° and 105°) ✓.

> **(ii)** The molecule has two bond pairs and two lone pairs. ✓ The repulsion between lone pairs is stronger than that between lone pair and bond pair, which is stronger than bond pair/bond pair repulsion. ✓ So the bond angle is reduced from the tetrahedral/109.5° bond angle to 104.5° ✓
>
> **(b)** C ✓

ⓔ The nitrogen atom in ammonia has three bond pairs and one lone pair. The basis of its shape is the (nearly) tetrahedral arrangement of these electron pairs, but the name of its shape is defined by the atom centres, and is pyramidal.

> **(c)** The attraction of a bonded atom for the electron pair in the bond. ✓
>
> **(d) (i)** The electronegativity of chlorine is greater than that of carbon, so electrons are unequally shared. ✓
>
> **(ii)** The molecule is tetrahedral and therefore symmetrical ✓ so the bond polarities cancel and the molecule is not polar ✓.

ⓔ Do *not* say that the charges cancel.

ⓔ **Total score: 10/10 marks**

Question 6

(a) The definition of the mole is:

  A the number of atoms in exactly 12 grams of the isotope $^{12}_{6}C$

  B the number of molecules in $24\,dm^3$ of a gas at $273\,K$ and $100\,kPa$

  C the number of molecules in $24\,dm^3$ of a gas at room temperature and $100\,kPa$

  D the amount of any substance containing the same number of atoms, molecules or groups of ions as there are atoms in exactly 12 grams of the isotope $^{12}_{6}C$

  (1 mark)

(b) Lithium chloride, LiCl, can be made by the reaction of lithium with chlorine:

$$2Li(s) + Cl_2(g) \rightarrow 2LiCl(s)$$

> **(i)** Calculate the maximum mass of lithium chloride that can be made from 35 g of lithium.

  (2 marks)

ⓔ The route is: mass of lithium → moles of lithium → moles of lithium chloride → mass of lithium chloride

(ii) Calculate the concentration in $mol\,dm^{-3}$ of the solution that would be obtained if this mass of lithium chloride were dissolved in water to make $5.00\,dm^3$ of solution. (1 mark)

(iii) Calculate the volume of chlorine gas required to react with 35 g of lithium. The molar volume of a gas at the temperature and pressure of the experiment is $24\,dm^3$. (2 marks)

e The route is: moles of lithium → moles of chlorine → volume of chlorine

(c) Explain why lithium has a higher melting temperature than sodium. (3 marks)

e You must understand metallic bonding to be able to answer this question. What are the forces involved and why are they stronger in one metal than in the other?

(d) (i) Define the term 'first ionisation energy'. (3 marks)

e You must state the amounts and the physical states (or state symbols if an equation is used) of all species.

(ii) Explain why the first ionisation energy of lithium is less endothermic than the first ionisation energy of neon. (3 marks)

(iii) Explain why the first ionisation energy of sodium is less endothermic than that of lithium. (3 marks)

(Total: 18 marks)

e You must compare the nuclear charges, the amounts of screening and the sizes of each.

Student answer

(a) D ✓

e Option A is simply a statement of the value of the Avogadro constant.

(b)

(i) moles of lithium used = $\dfrac{35\,g}{7.0\,g\,mol^{-1}}$ = 5.0 mol ✓ = moles of LiCl

mass of LiCl produced = 5.0 mol × 42.5 g mol^{-1} = 212.5 g ✓

(ii) concentration = $\dfrac{5.0\,mol}{5.00\,dm^3}$ = 1.0 mol dm^{-3} ✓

(iii) 1 mol Li reacts with 0.5 mol Cl_2 ✓

So:

volume Cl_2 required = 0.5 × 5.0 mol × 24 dm^3 mol^{-1} = 60 dm^3 ✓

e Note that the use of units throughout the calculations enables you to check that what you are doing is correct, since the units of the answer will also be correct.

(c) The lithium ion ✓ is smaller ✓ than the sodium ion, so the attraction between it and the delocalised electrons is greater ✓, making it harder to melt.

ⓔ Make sure that your answer refers to *both* sodium and lithium. Use of the word 'atom' for ion would lose 1 mark.

(d) (i) The energy change ✓ for the formation of a mole of gaseous unipositive ions ✓ from a mole of gaseous atoms ✓.

or

The energy change per mole ✓ for the process:

$E(g) \rightarrow E^+(g) + e^-$ ✓✓

ⓔ If you can define things through equations, do so. It is often quicker and clearer. The equation scores 2 marks because it includes both the positive ion and the gas phase.

Note that for chlorine it is the energy change for $Cl(g) \rightarrow Cl^+(g) + e^-$ and *not* for $\frac{1}{2}Cl_2(g) \rightarrow Cl^+(g) + e^-$.

(ii) The lithium atom is larger than the neon atom ✓ and the nuclear charge of lithium is lower than that of neon ✓. Both have two inner/shielding electrons, so the single outer electron is more effectively shielded from the +3 nucleus by the two inner electrons in lithium, whereas the outer eight electrons in neon are less shielded from the +10 nucleus by the two inner electrons ✓.

(iii) Sodium is a larger atom than lithium. ✓ The greater nuclear charge in sodium is offset ✓ by the increased number of shells, so there are increased repulsions/is more shielding from the inner electrons ✓.

ⓔ The idea of repulsion from inner electrons is sometimes expressed as shielding of the nuclear charge from the outer electrons by the inner ones. Some students find the notion of repulsions easier to visualise.

ⓔ **Total score: 18/18 marks**

Question 7

(a) 1.48 g of concentrated sulfuric acid was carefully dissolved in 250 cm³ of distilled water and the solution thoroughly mixed. This solution was titrated with 25.0 cm³ portions of 0.101 mol dm⁻³ sodium hydroxide solution. The mean titre was 21.77 cm³.

(i) Write the equation for this reaction. (1 mark)

(ii) Calculate the number of moles of sodium hydroxide per titration. (1 mark)

ⓔ Remember: moles = volume × concentration

(iii) Calculate the number of moles of sulfuric acid in the mean titre and hence the number in the sample. (2 marks)

e Use the stoichiometry of the equation.

(iv) Calculate the mass of pure sulfuric acid in the sample and hence its percentage purity. (2 marks)

e Remember: $\text{moles} = \dfrac{\text{mass}}{\text{molar mass}}$

(b) State and explain the procedure the student should have followed when making the solution of sulfuric acid to ensure a more accurate result. (3 marks)

(c) The student carried out a different titration and obtained the titres in the table below.

Experiment	1	2	3	4
Titre/cm^3	22.05	21.30	21.05	21.50

The mean titre is:

A 21.175

B 21.4

C 21.40

D 21.475 (1 mark)

(d) Concentrated sulfuric acid cannot be used to prepare a primary standard solution because it:

A is a dibasic acid

B is dangerous

C absorbs moisture from the air

D contains some dissolved acid sulfur trioxide (1 mark)

(Total: 11 marks)

Student answer

(a) (i) $H_2SO_4 + 2NaOH \rightarrow Na_2SO_4 + 2H_2O$ ✓

(ii) moles = volume × concentration = $0.0250\,dm^3 \times 0.101\,mol\,dm^{-3}$ = $0.002525\,mol$ ✓

(iii) moles of acid in $21.77\,cm^3 = \frac{1}{2} \times$ moles of NaOH = $0.0012625\,mol$ ✓

moles in $250\,cm^3 = \dfrac{0.0012625 \times 250}{21.77} = 0.014498$ ✓

(iv) mass of acid = moles × molar mass = $0.014498 \times 98.1\,g\,mol^{-1} = 1.42\,g$ ✓

percentage $= \dfrac{1.42 \times 100}{1.48} = 95.9\%$ ✓

---

e Set out your calculation, stating what you are calculating. Keep all figures on your calculator and round up at the end.

(b) Mixing the acid and $250\,cm^3$ of water will not accurately give $250\,cm^3$ of solution. You should carefully dilute the concentrated acid, allow it to cool, and then pour it and the washings ✓ into a volumetric flask ✓. Finally make up to $250\,cm^3$ ✓ and mix.

(c) C ✓

ⓔ The difference between the highest and lowest titres used to calculate the mean titre must not be more than $0.2\,cm^3$. B is wrong because the mean titre should be expressed to at least 2 decimal places. D is the mean of *all* the titres, not just the concordant ones. A implied that 21.05 is concordant with 21.30.

(d) C ✓

ⓔ A primary standard must not give off or absorb (e.g. from the air) any substance.

ⓔ Total score: 11/11 marks

Question 8

(a) Define the term *disproportionation*.   (2 marks)

(b) MnO_4^{2-} ions disproportionate in acid solution according to the equation:

$xH^+(aq) + yMnO_4^{2-}(aq) \rightarrow zMnO_4^-(aq) + (y - z)MnO_2(s) + \frac{1}{2}xH_2O$

The values of x, y and z are:

	x	y	z
A	4	2	1
B	2	2	1
C	4	3	1
D	2	3	1

(1 mark)

(c) Ammonium ions and nitrate ions react according to the equation:

$NH_4^+ + NO_3^- \rightarrow N_2O + 2H_2O$

(i) The oxidation numbers shown by nitrogen in these three species are:

	NH_4^+	NO_3^-	N_2O
A	+3	−5	−1
B	−3	+5	+1
C	+4	−6	−2
D	−4	+6	+2

(1 mark)

(ii) Explain why this reaction is not disproportionation.   (1 mark)

(Total: 5 marks)

Student answer

(a) It is when an element in a single species ✓ is simultaneously oxidised and reduced ✓.

(b) C ✓

e The manganese changes from +6 in MnO_4^{2-}, rising by 1 to +7 in MnO_4^- and falling by 2 to +4 in MnO_2, so there must be twice as many MnO_4^- ions as MnO_2. Thus there must be $3MnO_4^{2-}$. The number of H^+ can then be worked out by charge balance or by the number of oxygen atoms.

(c) (i) B ✓

e Hydrogen is always +1 (except when bonded to a metal) and oxygen is always minus (except when bonded to fluorine).

(ii) There are two nitrogen species, not a single nitrogen species, on the left. ✓

e This type of reaction, where an element in two species is both oxidised and reduced to a single oxidation state, is sometimes called reverse disproportionation.

e **Total score: 5/5 marks**

Question 9

(a) State and explain the trend in the first ionisation energy for the elements of group 2, Mg to Ba. (3 marks)

e State the trend and explain it in terms of number of protons, shielding and radius.

(b) Radium occurs below barium in group 2. On the basis of the trends in the chemistry of group 2, which of the following statements is true?

  A radium sulfate is more soluble in water than magnesium sulfate

  B radium hydroxide is more soluble in water than calcium hydroxide

  C radium carbonate is more easily decomposed by heating than magnesium carbonate

  D radium carbonate is very soluble in water (1 mark)

(c) The thermal stability of the nitrates of group 2 increases from magnesium nitrate to barium nitrate.

  (i) Give the equation for the thermal decomposition of magnesium nitrate. (2 marks)

  (ii) Explain why magnesium nitrate is less thermally stable than barium nitrate. (3 marks)

e Give your answer in terms of the effect of the size of the cation on the anion. Make sure that you use the correct terms — atom and ion mean different things.

(d) (i) State the flame colour produced by barium ions. (1 mark)

  (ii) Explain how a flame colour such as that from barium is produced by processes within the atom or ion. (2 marks)

(Total: 12 marks)

e Ensure that the order of events leading to the flame colour is clearly stated.

Questions & Answers

Student answer

(a) The first ionisation energy becomes less endothermic *or* falls ✓ as the group is descended *or* as the atomic number of the element becomes larger ✓, since the outer electron is further from the nucleus and is increasingly shielded by *or* repelled by the inner electrons ✓.

(b) B ✓

(c) (i) $Mg(NO_3)_2 \rightarrow MgO + 2NO_2 + \frac{1}{2}O_2$ ✓✓

ⓔ 1 mark is for MgO and NO_2, and 1 mark is for correct balancing.

State symbols are not needed here because the question does not ask for them. If they are required, a mark would be allocated for giving them correctly. If you are not asked to give state symbols, it is better to leave them out. The equation with state symbols is: $Mg(NO_3)_2(s) \rightarrow MgO(s) + 2NO_2(g) + \frac{1}{2}O_2(g)$

(ii) The magnesium ion is smaller than the barium ion ✓, so magnesium ions have a higher charge density and so are more polarising (than barium ions) ✓ and they distort the electron distribution in the nitrate ion more effectively (and therefore magnesium nitrate decomposes at a lower temperature) ✓.

ⓔ It is not enough for the first 2 marks to say that the charge density of a magnesium ion is higher than that of a barium ion.

Be careful to write magnesium *ion* or barium *ion*. The statement that 'magnesium has a higher charge density than barium' is often seen but never credited.

(d) (i) (Apple) green ✓

ⓔ Some apples are indeed other colours, but if it is a green apple, that's the colour of the barium flame.

(ii) Heating the atom/ion in the flame promotes electrons to higher energy levels. ✓ Electrons falling back down emit their energy as light of a specific colour. ✓

ⓔ **Total score: 12/12 marks**

Question 10

(a) Seawater contains aqueous bromide ions. During the manufacture of bromine, seawater is treated with chlorine gas and the following reaction occurs:

$$2Br^- + Cl_2 \rightarrow Br_2 + 2Cl^-$$

(i) Explain the term *oxidation* in terms of electron transfer. (1 mark)

(ii) Explain the term *oxidising agent* in terms of electron transfer. (1 mark)

(iii) State which of the elements, chlorine or bromine, is the stronger oxidising agent.

Explain the importance of this in the extraction of bromine from seawater, as represented by the equation above. (2 marks)

ⓔ Don't forget to include in your explanation which species is oxidised and which does the oxidising.

(b) When sodium chlorate(I), NaClO, is heated, sodium chlorate(v) and sodium chloride are formed.

(i) Write the ionic equation for this reaction. (2 marks)

ⓔ The equation must *not* include any spectator ions.

(ii) What type of reaction is this? (1 mark)

(iii) The same type of reaction occurs when chlorine is passed into cold dilute sodium hydroxide solution. Which chlorine-containing ions are produced?

A Cl^-, ClO_3^-

B OCl^-, Cl^-

C ClO_3^-, ClO_4^-, Cl^-

D ClO_3^-, OCl^- (1 mark)

(c) When concentrated sulfuric acid is added to potassium bromide, orange-brown fumes of bromine and sulfur dioxide gas are produced.

(i) Write an equation to show the production of these two gases. (2 marks)

(ii) State the oxidation numbers of sulfur in SO_2 and SO_4^{2-}. (2 marks)

ⓔ Include a charge as well as the number.

(iii) Use your answer to part (c)(ii) to explain whether H_2SO_4 is oxidised or reduced in the above reaction. (1 mark)

(iv) When the gases produced are passed into a solution of potassium iodide, iodine is produced. Name a reagent that could be used to confirm that a solution contains iodine. State what you would see. (2 marks)

(Total: 15 marks)

Student answer

(a) (i) Oxidation is electron loss. ✓

(ii) An oxidising agent gains electrons/removes electrons from another substance. ✓

(iii) Chlorine is the stronger oxidising agent ✓; it removes electrons from bromide ions ✓ to produce bromine.

Questions & Answers

e It is important to distinguish between bromine and bromide. Always make clear which species is under consideration. Similarly, 'magnesium' is an element — if you mean 'magnesium ions' (Mg^{2+}), you must include the word 'ions'.

(b) (i) $3ClO^- \rightarrow 2Cl^- + ClO_3^-$ ✓✓

e There is 1 mark for the species and 1 mark for the balanced equation. Note that the question asks specifically for the ionic equation. The non-ionic version including the sodium might get 1 mark.

(ii) Disproportionation ✓

e It is true that this is also a redox reaction. However, it is a particular type of redox reaction and this is the more significant piece of information.

(iii) B ✓

e The ionic equation for the reaction is: $Cl_2 + 2OH^- \rightarrow Cl^- + OCl^- + H_2O$

The chlorine has disproportionated. The answer would be option A if the sodium hydroxide solution had been hot, since OCl^- ions disproportionate on heating to chloride ions and chlorate(v) ions.

(c)

(i) Either

$2NaBr + 2H_2SO_4 \rightarrow Na_2SO_4 + 2H_2O + SO_2 + Br_2$

or

$2HBr + H_2SO_4 \rightarrow 2H_2O + SO_2 + Br_2$ ✓✓

e There is 1 mark for the species and 1 mark for the balanced equation.

(ii) In SO_2 the oxidation number of sulfur is +4 ✓; in SO_4^{2-} it is +6 ✓.

e Note that there is no credit for showing any working because there is only 1 mark available for each answer. This is unlike other calculation-type questions in which there are often marks available for working, even if the answer is wrong.

(iii) Reduced, because the oxidation state of sulfur has decreased/gone from +6 to +4. ✓

(iv) Starch ✓; it turns blue-black ✓.

e Starch-iodide paper should not be used in this test; the iodide ions could be oxidised to iodine by either chlorine or bromine, so starch-iodide paper turns blue-black with all the halogens. Starch turns blue-black only with iodine.

e **Total score: 15/15 marks**

Question 11

Chlorine reacts in a very exothermic reaction with cyclohexene, C_6H_{10}.

$$Cl_2(g) + C_6H_{10}(l) \rightarrow C_6H_{10}Cl_2(l)$$

Data: molar volume of a gas under the experimental conditions = $24\,dm^3\,mol^{-1}$

density of liquid cyclohexene = $0.811\,g\,cm^{-3}$

boiling temperature of cyclohexene = $82°C$

(a) Calculate the volume of chlorine needed to react with $2.5\,cm^3$ of liquid cyclohexene.

(3 marks)

(b) A student carried out this preparation using the apparatus shown in the diagram.

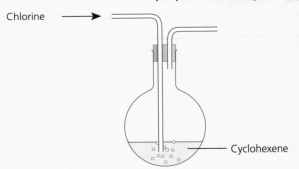

Chlorine ⟶

Cyclohexene

(i) Explain the difference between hazard and risk, using chlorine as an example.

(2 marks)

(ii) State the main hazard of this experiment and suggest how the associated risk should be reduced.

(2 marks)

(iii) Explain what alteration could be made to the experiment to ensure a higher yield of product.

(2 marks)

(Total: 9 marks)

Student answer

(a) mass of cyclohexene = volume × density = $2.5\,cm^3$ × $0.811\,g\,cm^{-3}$ = $2.0275\,g$ ✓

amount (moles) of cyclohexene = $\dfrac{mass}{molar\ mass}$ = $\dfrac{2.0275\,g}{82\,g\,mol^{-1}}$ = $0.024725\,mol$ =

moles of chlorine needed ✓

volume of chlorine gas = moles × molar volume = $0.024275\,mol$ × $24\,dm^3\,mol^{-1}$ = $0.593\,dm^3$ or $593\,cm^3$ ✓

---

e It is a good idea to add units in the calculation of each step. This way you can see whether you had the expression, for example, mass/molar mass the right way up.

Questions & Answers

(b) (i) A hazard is the intrinsic property of a chemical ✓, whereas risk refers to how the hazardous substance is used ✓.

(ii) Chlorine is toxic/an irritant ✓ so, to reduce the risk, the experiment should be performed in a fume cupboard ✓.

e Wearing safety glasses and a lab coat are not regarded as special safety features — their use is standard practice.

(iii) If the temperature got too high some of the cyclohexene would vaporise and so reduce the yield ✓. To avoid this, the flask should be placed in a beaker of iced water ✓.

e Total score: 9/9 marks

Question 12

(a) A solution of benzene diazonium chloride, $C_6H_5N_2Cl$, was weighed and then gently warmed. The gas evolved was collected and then the solution was reweighed.

Data: mass of solution before reaction = 28.39 g

mass of solution after reaction = 28.12 g

volume of gas produced = 237 cm³

pressure of gas = 97.0 kPa

temperature of gas = 18.5°C

the gas constant, R, = 8.31 J K^{-1} mol^{-1}

(i) State the gas law. (1 mark)

(ii) Use it to calculate the number of moles of gas produced and hence suggest its molar mass. (5 marks)

(iii) Use your answer to (ii) to suggest the identity of the gas. (1 mark)

(b) Benzene diazonium chloride solution contains chloride ions. When a solution containing 2.81 g of $C_6H_5N_2Cl$ was added to excess silver nitrate solution, 2.75 g of solid silver chloride, AgCl, was obtained. Calculate:

(i) the theoretical mass of silver chloride that should be produced. (4 marks)

(ii) the percentage yield of the precipitation reaction. (1 mark)

(Total: 12 marks)

Student answer

(a) (i) $pV = nRT$ ✓

(ii) p = 97.0 kPa = 97 000 Pa ✓

V = 237 cm³ = 0.237 dm³ = 2.37 × 10^{-4} m³ ✓

T = 18.5 + 273 = 291.5 K

ⓔ Note that pressures must be converted to Pa, volumes to m^3 and the temperature to kelvin.

> mass of gas produced = 28.39 − 28.12 = 0.27 g ✓
>
> $n = \dfrac{97\,000 \times 2.37 \times 10^{-4}}{8.31 \times 291.5} = 0.00949$ ✓
>
> molar mass $= \dfrac{\text{mass}}{\text{moles}} = \dfrac{0.27}{0.00949} = 28.5\,\text{g mol}^{-1}$ ✓
>
> **(iii)** The gas is likely to be nitrogen, molar mass $28\,\text{g mol}^{-1}$. ✓
>
> **(b) (i)** molar mass of $C_6H_5N_2Cl = 140.5\,\text{g mol}^{-1}$ ✓
>
> moles of $C_6H_5N_2Cl = \dfrac{2.81}{140.5} = 0.0200$
>
> = theoretical moles of AgCl ✓
>
> molar mass of AgCl = 107.9 + 35.5 = $143.4\,\text{g mol}^{-1}$ ✓
>
> theoretical mass of AgCl = 0.0200 × 143.4 = 2.87 g ✓

ⓔ The actual mass obtained (2.75 g) is not used until the final calculation of percentage yield.

> **(ii)** percentage yield $= \dfrac{\text{actual mass}}{\text{theoretical mass}} \times 100 = \dfrac{2.75}{2.87} \times 100 = 95.8\%$ ✓

ⓔ Total score: 12/12 marks

Knowledge check answers

1 number of protons in both isotopes = 35; number of neutrons in ^{79}Br = 79 − 35 = 44; number of neutrons in ^{81}Br = 81 − 35 = 46

2 formula mass of $Ca(OH)_2$ = 40.1 + 2 × (16 + 1) = 74.1 (no units)

3 It will have three peaks. One at m/z = 158, due to $(^{79}Br-^{79}Br)^+$, one at m/z = 160, due to $(^{79}Br-^{81}Br)^+$ and one at m/z = 162, due to $(^{81}Br-^{81}Br)^+$.

4 A_r = (35 × 0.755) + (37 × 0.245) = 35.49

5 **(a)** Si: $1s^2\ 2s^2\ 2p^6\ 3s^2\ 3p^2$
(b) Mn: $1s^2\ 2s^2\ 2p^6\ 3s^2\ 3p^6\ 4s^2\ 3d^5$

6 [Ar] 4s ⊞ 3d ⊞⊞⊞⊞⊞

7 **(a)**
$$\left[\begin{array}{c} \overset{\times\times}{\underset{\times\times}{\times\ Mg\ \times}} \end{array} \right]^{2+}$$
(b)
$$\left[\begin{array}{c} \overset{\times\times}{\underset{\times\times}{\times\times\ Cl\ \times\times}} \end{array} \right]^{-}$$

8 Both ions have ten electrons but the F^- ion has nine protons, whereas the O^{2-} ion has eight. This means that there is less attraction in the O^{2-} ion between the nucleus and the outer electrons.

9 ten

10

11 $:\overset{..}{\underset{..}{Cl}}\ \overset{\times\times}{\underset{\times\times}{P}}\ \overset{..}{\underset{..}{Cl}}:$
$\overset{..}{\underset{..}{Cl}}:$

12 Magnesium has more delocalised electrons and a smaller ionic radius than sodium, so the forces between the ions and the delocalised electrons are stronger.

13 two

14 Pyramidal — nitrogen has three bond pairs and one lone pair).

15 OF_2. The O in OF_2 has two lone pairs. The P in PH_3 has one lone pair and the Si in SiH_4 has no lone pairs. The lone pair/lone pair repulsion squashes the F–O bond more than the lone pair/bond pair repulsion in PH_3 or the bond pair/bond pair in SiH_4.

16 One lone pair (and three bond pairs), so it is tetrahedral.

17 All contain polar bonds except CI_4, where both C and I have electronegativities of 2.5.

18 Only OF_2 and CH_2Cl_2. In the others with polar bonds, the bond polarities cancel, as the molecules are symmetrical.

19 London (dispersion) and dipole–dipole forces

20 London (dispersion) force

21 Ethanol. It has an O–H group and so can form hydrogen bonds with water. These are similar in strength to the hydrogen bonds in water. Dimethyl ether has no δ+ hydrogen and so cannot form hydrogen bonds.

22 Although propanone cannot form hydrogen bonds with itself (it has no δ+ hydrogen atom), its δ− oxygen can form hydrogen bonds with the δ+ hydrogen atoms in water.

23 The most soluble in water is $C_2H_5NH_2$ (it can hydrogen bond with the water). In non-polar hexane it is the non-polar C_2H_6.

24 There are strong electrostatic forces between the oppositely charged ions. A lot of energy is therefore needed to separate the ions.

25 $Cr_2O_7^{2-}$: the seven oxygen atoms add up to −14. The ion is −2, so the two chromium atoms add up to +12, or +6 each.
NH_4^+: the four hydrogen atoms add up to +4, the ion is +1, so the nitrogen is −3.

26 The tin in $SnCl_2$ has been oxidised as its oxidation number has increased from +2 to +4. The iron in $FeCl_3$ has been reduced as its oxidation number has decreased from +3 to +2.

27 $Li_2CO_3 \rightarrow Li_2O + CO_2$

28 The flame test indicates calcium. The single acidic gas is CO_2, so the compound is calcium carbonate. (Calcium nitrate would give two gases.)

29 number of methane molecules = 0.125 mol × 6.02 × 10^{23} mol^{-1} = 7.53 × 10^{22}

30 total number of ions = 5 × 0.125 mol × 6.02 × 10^{23} mol^{-1}
= 3.76 × 10^{23}

31 **(a)** 12 + 4 = 16 g
(b) 23.0 + 35.5 = 58.5 g

32 **(a)** moles = $\dfrac{\text{mass}}{\text{molar mass}}$ = $\dfrac{2.0\,g}{16\,g\,mol^{-1}}$ = 0.125 mol
(b) moles = $\dfrac{2.0\,g}{58.5\,g\,mol^{-1}}$ = 0.034 mol

33 mass = moles × molar mass = 0.12 mol × 40 g mol^{-1} = 4.8 g

34 $2Al + 6HCl \rightarrow 2AlCl_3 + 3H_2$

35 $Sn^{2+} + 2Fe^{3+} \rightarrow Sn^{4+} + 2Fe^{2+}$

36 volume = moles × molar volume = 0.0246 mol × 24 dm^3 mol^{-1} = 0.59 dm^3

37 volume of solution = $\dfrac{250}{1000}$ = 0.250 dm^3
moles of NaOH needed = concentration × volume in dm^3
= 0.100 mol dm^{-3} × 0.250 dm^3 = 0.0250 mol
mass of NaOH needed = moles × molar mass
= 0.0250 mol × 40 g mol^{-1} = 1.00 g

38 weighing error = $\dfrac{2 \times 0.01 \times 100}{1.23}$ = 1.63%
pipette error = $\dfrac{0.1 \times 100}{25.0}$ = 0.40%
burette error = $\dfrac{2 \times 0.05 \times 100}{21.40}$ = 0.47%
total error = 1.63 + 0.40 + 0.47 = 2.50%

39 amount of ethanedioic acid = $0.0500 \, mol \, dm^{-3} \times 0.025 \, dm^3 = 0.00125 \, mol$

amount of sodium hydroxide = $2 \times 0.00125 = 0.0025 \, mol$

concentration of sodium hydroxide = $\dfrac{moles}{volume} = \dfrac{0.0025 \, mol}{0.0234 \, dm^3} = 0.107 \, mol \, dm^{-3}$ (to 3 significant figures)

40 $CH_3COOH + C_2H_5OH \rightarrow CH_3COOC_2H_5 + H_2O$ has the higher atom economy as the sum of the molar masses of the reactants is less than that in the other equation.

41 molar mass of $H_2C_2O_4.2H_2O = 2 + (2 \times 12) + (4 \times 16) + (2 \times 18) = 126 \, g \, mol^{-1}$

moles required = $\frac{1}{4} \times 0.0500 = 0.0125$

mass required = $126 \, g \, mol^{-1} \times 0.0125 \, mol = 1.575 \, g$

Index

A

addition reactions   55
alcohols   25, 26
alkaline earth metals (group 2)   13, 32–35
alkanes   24–25
aluminium, ionisation energies   11
aluminium oxide   16
ammonia   18, 19, 21, 39
ammonium ion   19, 41
amount, as a term   42
analogous molecules   18
anions   14, 41
assessment objectives   59–60
astatine   40
atom economies   55
atomic mass   10
atomic number   8
atoms, definition   8
Avogadro constant   42

B

Balmer series   11
barium   35
beryllium chloride   19
boiling temperatures   13–14, 16, 23–25
bonding
  covalent bonding   16–17
  intermediate bonding   21–22
  ionic bonding   14–16
  metallic bonding   17
bond polarity   21–22
boron trichloride   19
boron trifluoride   21
branched-chain isomers   25
bromine   36, 37
bromothymol blue   53
buckminsterfullerene   28
burettes   52–53
butane   24, 25

C

caesium   21
calcium   35
calcium carbonate   35

calculations
  gas volumes   47–51
  general   6–7
carbon   13
carbonate ion   20
carbonates
  analysis   41
  of groups 1 and 2   35
carbon atoms   28
carbon dating   10
carbon dioxide   20
carbon nanotubes   28
carbon tetrachloride   21
cations   14, 15, 41
chemical equations   6, 45–46
chloride   15
chlorine   33, 36, 37, 38
chromium   13
command terms   60
concentration   51–52
copper   13
copper chromate   16
core practicals
  hydrochloric acid solution   57
  molar volume of a gas   48
  standard solution   56–57
Coulomb's law   27
covalent bonding   16–17
covalent bonds   12, 13, 16–17, 21

D

dative covalent bonds   16
d-block   11
diagrams   6
diamond   28
dimethylpropane   25
disproportionation
  definition   31
  group 7 (halogens)   38
d-orbitals   13
dot-and-cross diagrams
  covalent molecules   17
  ions   14–15
double bonds   20
Drude–Lorentz model   17

E

electrolysis   15–16
electron configuration   11–13
electronegativity   21
electron orbits   11
electrons, properties   8
electron transfer   30
elimination reactions   55
empirical formulae   43–44
enthalpy   25
equations
  chemical   45–46
  gas law   47
  ionic   45–46
  reacting masses   46–47
errors   53
ethane   24, 25
ethanol, mass spectrum of   9
exam papers   59–62

F

first ionisation energy   10–11
flame tests, groups 1 and 2   35
fluorine   21, 24, 36, 40
formulae   42–44, 43–44

G

gas law   47
gas volumes, calculations   47–51
germane   24
giant covalent lattices   28
giant ionic lattices   27
graphene   28
graphite   28
graphs   6
ground state   11
group 2 (alkaline earth metals)   13, 32–35
group 4 hydrides   24
group 5 hydrides   24
group 7 (halogens)   36–40

H

half-equations   31–32
halides   41
halide salts   38–39

halogens (group 7)   36–40
hazard   58
hexane   24
H–O–H angles   18
hydrides
    of group 4   24
    of group 5   24
hydrocarbons   26
hydrochloric acid solution, core
        practical   57
hydrogen
    electron configuration   11
    energising   11
    oxidation number   29
hydrogen bonds   22–23
hydrogencarbonates   41
hydrogen chloride   19
hydrogen halides   39
hydrogen sulfide   18

I
ice   27
indicators   53
inorganic compounds,
        analysis   41
instantaneous dipole–induced
        dipole forces   23
intermediate bonding   21–22
intermolecular forces
        16, 22–23
iodine   27, 36
ionic bonding   14–16
ionic bonds   21, 27
ionic compounds
    polarity   21
    solubility   25
ionic equations   45–46
ionic radii   15
ionisation energies   10–12, 32
ions
    evidence of existence   15–16
    formation   14–15
    shapes of   18–20
iron   37
isoelectronic ions   15
isotopes, definition   8

L
learning techniques   5–6
Lewis structure   17
lithium   35
lithium carbonate   35
lithium nitrate   35
London forces   23, 25
Lyman series   11

M
magnesium   30, 35
magnesium chloride   21
magnesium iodide   21
mass number   8
mass spectral data, uses of
        9–10
mass spectrometers   9–10
materials, properties of   23–25
melting temperatures   13–14, 16,
        17, 23–24, 27, 29, 36
metal ions   15
metallic bonding   17
metals, structure   28
methane   18, 19, 24
methylbutane   25
methyl orange   53
molar mass   43
    of volatile liquids   47–48
molar volume, of a gas   48–51
molecular formulae   43–44
molecules
    polarity   21
    shapes   18–20
moles   42
multiple-choice questions   61

N
negative ions   14
neon   9
neutrons, properties   8
nitrate ion   20
nitrates, groups 1 and 2   34–35
nitrogen   24
noble (inert) gases   13, 23
non-metal ions   15
note making   7

O
online marking   61–62
orbitals   12–13, 16
oxidation numbers   29–31
oxidising agents   37
oxidising reactions, group 7
        (halogens)   36–37
oxygen
    boiling temperature   24
    oxidation number   29
    reactions with group 2
        metals   33

P
p-electrons   11
pentane   24, 25
percentage yields   54–55
periodic properties   13–14
periodic trends   13–14
permanent dipole–permanent
        dipole forces   23
phenolphthalein   53
phosphine   18
phosphorus   12
phosphorus pentachloride   19
physical properties
    group 7 (halogens)   36
    and structure   29
pipettes   52
pitfalls, exams   62
planning   5
polar bonds   21–22
polar covalent bonds   21
polarity   21–22
polar molecules   23
    in water   26
polyatomic anions   15
p-orbitals   12–13
positive ions   14
potassium   35
potassium iodide   36
practical chemistry, questions   48,
        56–58
propane   24, 25
properties of materials   23–25
protons, properties   8

Index

Q
quantum shells   11

R
radioactive dating   10
reacting masses, equations   46–47
reactions
   atom economies   55
   group 2 metals   33
   group 2 oxides   33–34
   group 7 (halogens)   36–37
redox   29–32
relative atomic mass   8
relative formula mass   8
relative isotope mass   8
relative molecular mass   8, 42
revision   60
risk   58

S
second ionisation energy   10
s-electrons   11
shapes
   of ions   18–20
   of molecules   18–20
silane   18, 24
silicon   13

silver nitrate   39
simple molecular structures   27
single bonds   18–19
sodium   15, 28, 35
sodium chloride   27
solids
   boiling and melting
     temperatures   23
   structure   26–29
solubility   25–26
   group 2 hydroxides   34
s-orbitals   12–13
spectator ions   45
spectral evidence, electron
   configuration   11
standard solutions   56–57
stannane   24
state symbols   45
stoichiometry   29
strontium   35
structure, solids   26–29
substitution reactions   55
sulfate ion   20
sulfates   41
sulfur dioxide   20
sulfur hexafluoride   19
sulfuric acid   38–39

T
tables   6
targets   6
textbooks   6
thermal stability, group 2
   metals   34–35
titration calculations   51–54

V
valence-shell electron-pair repulsion
   theory   18–20
vanadium   12
van der Waals forces   23
volatile liquids, molar mass of
   47–48
volumetric analysis   51–54

W
water
   H–O–H angle   18
   hydrogen bonds   22–23
   and hydrogen halides   39
   reactions with group 2
     metals   33
   reactions with group 2 oxides   33
   relative molecular mass   42
   shape and structure   19
writing skills   7

The periodic table

Key:

Relative atomic mass
Atomic symbol
name
Atomic (proton) number

Group

Period	1	2												3	4	5	6	7	0
1	1.0 **H** hydrogen 1																		4.0 **He** helium 2
2	6.9 **Li** lithium 3	9.0 **Be** beryllium 4												10.8 **B** boron 5	12.0 **C** carbon 6	14.0 **N** nitrogen 7	16.0 **O** oxygen 8	19.0 **F** fluorine 9	20.2 **Ne** neon 10
3	23.0 **Na** sodium 11	24.3 **Mg** magnesium 12												27.0 **Al** aluminium 13	28.1 **Si** silicon 14	31.0 **P** phosphorus 15	32.1 **S** sulfur 16	35.5 **Cl** chlorine 17	39.9 **Ar** argon 18
4	39.1 **K** potassium 19	40.1 **Ca** calcium 20	45.0 **Sc** scandium 21	47.9 **Ti** titanium 22	50.9 **V** vanadium 23	52.0 **Cr** chromium 24	54.9 **Mn** manganese 25	55.8 **Fe** iron 26	58.9 **Co** cobalt 27	58.7 **Ni** nickel 28	63.5 **Cu** copper 29	65.4 **Zn** zinc 30		69.7 **Ga** gallium 31	72.6 **Ge** germanium 32	74.9 **As** arsenic 33	79.0 **Se** selenium 34	79.9 **Br** bromine 35	83.8 **Kr** krypton 36
5	85.5 **Rb** rubidium 37	87.6 **Sr** strontium 38	88.9 **Y** yttrium 39	91.2 **Zr** zirconium 40	92.9 **Nb** niobium 41	95.9 **Mo** molybdenum 42	[98] **Tc** technetium 43	101.1 **Ru** ruthenium 44	102.9 **Rh** rhodium 45	106.4 **Pd** palladium 46	107.9 **Ag** silver 47	112.4 **Cd** cadmium 48		114.8 **In** indium 49	118.7 **Sn** tin 50	121.8 **Sb** antimony 51	127.6 **Te** tellurium 52	126.9 **I** iodine 53	131.3 **Xe** xenon 54
6	132.9 **Cs** caesium 55	137.3 **Ba** barium 56	138.9 **La** lanthanum 57	178.5 **Hf** hafnium 72	180.9 **Ta** tantalum 73	183.8 **W** tungsten 74	186.2 **Re** rhenium 75	190.2 **Os** osmium 76	192.2 **Ir** iridium 77	195.1 **Pt** platinum 78	197.0 **Au** gold 79	200.6 **Hg** mercury 80		204.4 **Tl** thallium 81	207.2 **Pb** lead 82	209.0 **Bi** bismuth 83	[209] **Po** polonium 84	[210] **At** astatine 85	[222] **Rn** radon 86
7	[223] **Fr** francium 87	[226] **Ra** radium 88	[227] **Ac** actinium 89	[261] **Rf** rutherfordium 104	[262] **Db** dubnium 105	[266] **Sg** seaborgium 106	[264] **Bh** bohrium 107	[277] **Hs** hassium 108	[268] **Mt** meitnerium 109	[271] **Ds** darmstadtium 110	[272] **Rg** roentgenium 111								

Elements with atomic numbers 112–116 have been reported but not fully authenticated

140.1 **Ce** cerium 58	140.9 **Pr** praseodymium 59	144.2 **Nd** neodymium 60	144.9 **Pm** promethium 61	150.4 **Sm** samarium 62	152.0 **Eu** europium 63	157.2 **Gd** gadolinium 64	158.9 **Tb** terbium 65	162.5 **Dy** dysprosium 66	164.9 **Ho** holmium 67	167.3 **Er** erbium 68	168.9 **Tm** thulium 69	173.0 **Yb** ytterbium 70	175.0 **Lu** lutetium 71
232 **Th** thorium 90	[231] **Pa** protactinium 91	238.1 **U** uranium 92	[237] **Np** neptunium 93	[242] **Pu** plutonium 94	[243] **Am** americium 95	[247] **Cm** curium 96	[245] **Bk** berkelium 97	[251] **Cf** californium 98	[254] **Es** einsteinium 99	[253] **Fm** fermium 100	[256] **Md** mendelevium 101	[254] **No** nobelium 102	[257] **Lr** lawrencium 103